DO FISH FEEL PAIN?

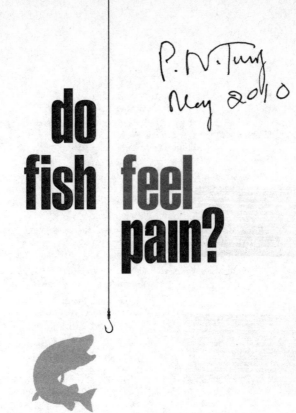

P. N. Tung
May 2010

do fish feel pain?

VICTORIA BRAITHWAITE

OXFORD
UNIVERSITY PRESS

OXFORD
UNIVERSITY PRESS

Great Clarendon Street, Oxford OX2 6DP

Oxford University Press is a department of the University of Oxford.
It furthers the University's objective of excellence in research, scholarship,
and education by publishing worldwide in

Oxford New York

Auckland Cape Town Dar es Salaam Hong Kong Karachi
Kuala Lumpur Madrid Melbourne Mexico City Nairobi
New Delhi Shanghai Taipei Toronto

With offices in

Argentina Austria Brazil Chile Czech Republic France Greece
Guatemala Hungary Italy Japan Poland Portugal Singapore
South Korea Switzerland Thailand Turkey Ukraine Vietnam

Oxford is a registered trade mark of Oxford University Press
in the UK and in certain other countries

Published in the United States
by Oxford University Press Inc., New York

British Library Cataloguing in Publication Data

Data available

Library of Congress Cataloging in Publication Data

Data available

Typeset by SPI Publisher Services, Pondicherry, India
Printed in Great Britain
on acid-free paper by
Clays Ltd., St Ives Plc

ISBN 978–0–19–955120–0

1 3 5 7 9 10 8 6 4 2

FOR

Andrew, James and Matthew

Preface

In 2006 Nick Goldberg, an editor at the *Los Angeles Times*, asked me to write a brief Op-Ed piece on whether fish feel pain. After the article appeared, the newspaper and I received letters and emails. These were of two sorts. Some told me that I was persecuting anglers by spreading untruths and myths—wasn't it clear to everyone that fish don't feel pain? But the others wanted to know why I bothered to investigate the question—wasn't it clear to everyone that fish do feel pain?

I had a certain amount of sympathy with both camps. I could identify with those who believed I was threatening the angling community. That was not my intention, but there had been a great deal of inaccurate information written about research on pain in fish so it was understandable that some people were being defensive. On the other hand, how were the others to know that no scientific analysis of even the basics of fish pain had been conducted before the turn of *this* century?

Those polarized reactions, which also played out on various websites, prompted me to wonder whether there was a need for a fuller account of the science behind the fish pain debate. The result is this book.

I haven't always been a fish biologist. I started my research career working with birds, asking questions on cognition such as 'What makes some animals smarter than others?' Fifteen years ago, however, I switched to fish. To me it wasn't a big change, I was still asking the same kind of question, but it was easier to compare cognition among different populations of fish than it was for birds. To many of my colleagues, though, it was a curious move, and several even thought it a move backwards. 'Why fish?' they would ask me, and invariably this was followed by a bit of a snigger and, 'Don't they have a three second memory?' The reaction of my colleagues was telling—fish are perceived as less worthy. But why?

Up to then, my experience with fish was very limited. Like many children, I kept goldfish when I was younger, but other than that I knew very little. Yet as I discovered more and more about the biology, physiology and behaviour of fish, I became engrossed. They really are seductive. My family know this to their cost because I can rarely pass a pond, stream or river without stopping to search for a tiny bit of movement, the slightest flash of silver that betrays a fish's position. It has sometimes seemed that traditional roles have been reversed in my family—more than once one of my young sons would slip his hand into mine and plead, 'Come on Mum!', as he coaxed me away from the water's edge.

To this day, however, I don't regret my decision to move from feathered creatures to venture underwater into a piscine world. Fish are smart if you ask the right questions. And, by the way, it turns out that several fish species have excellent memories that can last several days, and in some cases even months.

My goal in writing this book has been to provide the background to promote informed discussion. Like other animal welfare debates, constructively arguing about fish welfare requires that we understand the issues, that we review evidence and discuss this appropriately. In the book, I examine what we know so far about pain in fish, and whether it is meaningful to discuss fish welfare at all. After reading the book, I hope you will be in a position to make up your own mind. I have no axe to grind—I choose to eat fish and I experiment on them, but while I have been fishing in the past, I am not an active angler though I have many friends and colleagues who are. As the book began to take shape it became clear that the fish pain debate probes questions about science, welfare and ethics. It draws us towards difficult, grey areas—if fish feel pain, then what about octopus, squid and lobsters—where do we draw the line? This might be the first book in a series, or the next one might be the last.

Much of the material I present has benefited from discussions with many colleagues and friends. As the book began I was lucky enough to be resident at the Wissenshcaftskolleg zu Berlin (Institute for Advanced Study) in

Germany. I could not have asked for a more stimulating place to think and write and I thank Wiko and the many Fellows who took time to discuss pain in fish with me. When I first began to focus on fish welfare I found conversations with my long term mentor, collaborator and friend Felicity Huntingford incredibly useful—she and the colleagues she introduced me to helped shape my views. Throughout the writing of this book I have had collaborations with the University of Bergen and the Institute for Marine Research in Norway. Many people there have shared their opinions and answered questions, but in particular I thank my collaborator and friend Anne Gro Vea Salvanes and our student Olav Moberg for their continuing input. And I am grateful to Mike Gentle for first suggesting that we get together to do our part in the science I describe in the book and to the UK's Biotechnology and Biological Sciences Research Council for funding it. I also thank Bob Elwood for constructively disagreeing with my views on hermit crabs. My new Penn State University colleagues especially Bob Carline and Gary San Julian have been stimulating foils for debate. I hope they see the merit in discussing this. Even if we don't discuss it, others will.

As the concept of the book was forming I had doubts and I am grateful to Gabrielle Archard, Mike Beentjes, Phil Boulcott, Nichola Brydges, Zach Colvin, Clive Copeman, Bryan Ferguson, Cairsty Grassie, Sue Healy, Andrew Illius, In Kim, Sean Nee, Mark Viney, Dan Weary and my New

Jersey-Yorkshire family, especially Jo, Cathy and Sam, for encouraging me on. As the book was drawing to a close my first mentor, Marian Dawkins, provided advice on chapter 4. Much of how I think about animal welfare comes from time spent with Marian. Again and again I am amazed at how far ahead of her time she has been and how articulately she explains the welfare world. Her impact on welfare science has been substantial and is likely to become greater as the scientific community catches up.

Two people have been instrumental in getting this project through to completion. Latha Menon is a wonderful editor, and I thank her for her patience and vision, and for offering me the opportunity to write a book in the first place. To Andrew Read I owe an enormous debt of gratitude. It isn't easy living with someone when they are writing a focused piece of work and Andrew has put up with me doing this twice, once as I wrote and then published my PhD thesis and now as I have written this book. Andrew has been a sharp-eyed critic and an interested audience, and at the same time the most supportive partner anyone could hope for. Any errors are his!

University Park, Pennsylvania,
October 2009

Contents

The Problem

In 2003 the results from a study investigating whether fish feel pain were published. Almost overnight the research article captured the media's attention, and the authors found themselves propelled into the limelight. They were asked to appear on live radio and television and invited to speak to journalists from around the world. The findings had made front page news. The issue of fish pain seemed to resonate for many people. After the phones stopped ringing and the dust had settled, the initial frenzy turned to reflection. The scientific debate about fish pain was underway. A few years on, and the discussions persist. And among nonscientists, many people firmly believe fish are dim-witted creatures incapable of feeling pain. But others, equally committed to their beliefs, argue that we should provide fish with the same level of care and welfare that we do for birds and mammals. So who is right—and does it really matter?

As one of the authors of the original research article, I continue to be amazed by the interest that this topic has generated. We even wound up in a passage in a best-selling novel.[1] A great deal has now been said and written; every few months summaries of the debate crop up in newspapers, and cyberspace chat rooms continue to fill with discourse and disagreement. It is clear from this outpouring that strong feelings fuel the fish pain debate, but also that the discussions are based on both fact and fiction—so much so that it has become difficult to distinguish between the two. This hasn't been helped by the fact that much of the scientific material underpinning the debate is buried in research papers in technical journals. The aim of this book is to bring the science behind the debate into the open—I have no personal agenda here other than to make the facts and the reasoning more accessible.

Asking whether fish feel pain piques the interest of a truly eclectic group of people: from anglers to scientists, from aquarium enthusiasts to ethicists, and from welfare campaigners to legislators. Accepting that an animal has the ability to suffer from pain changes the way we choose to interact, handle, and care for it. Knowing that something

---

[1] Ian McEwan, *Saturday* (London: Vintage, 2006), 127 ff.

might suffer in our hands influences the moral and ethical judgments that we make. We become concerned for the animal.

Pain is a negative, unpleasant sensation that we try to avoid. It makes most of us feel uncomfortable to know that someone else is hurting. That is also the case where the someone else is an animal that we can relate to, such as a monkey or a dog. Our ability to empathize with another's suffering seems to be part of human nature, but when we direct this empathy towards an animal rather than a person, is this some strange misdirected anthropomorphism or is it appropriate for us to show such concern? We find it difficult to make distinctions about whom or what we should care for and protect—this is why we *debate* the possibility of pain in fish. While it is readily accepted that we should protect another human being, even newborn babies with a still-developing nervous system, the clarity of that decision begins to wane when we consider how to respond to an injured animal. There is without doubt a considerable distance between wanting to protect and alleviate pain in a person and wanting to do so in a fish. But the curious thing is that as we try to explain why there is this gap we begin to stray into uncomfortable territory where there are more questions than answers—a pall of uncertainty descends.

It seems strange that it has taken until now for us to ask whether fish feel pain. Is it because we think we know already, or because we don't like to think about the consequences of concluding that they suffer? Or is it because it's a very hard question to answer? That it certainly is: the question challenges both scientific and philosophical ideas and it forces us to think about pain as a mechanism— what it is and how it works. When we are in pain it hurts—we suffer. Do other animals share this ability to experience negative feelings? Several researchers argue that feelings and emotions are exclusive to humans and dismiss the idea that animals can suffer. Yet we accord pets and farm animals welfare rights. Asking if fish feel pain challenges established ideas; it is akin to opening the proverbial can of worms—as we pose the question, a whole slew of unknowns arise. Which animals should we care about from an ethical point of view? Are fish conscious? Where should we draw the line? Should fish be on the same side as birds and mammals, or should they be categorized alongside lobsters, squid, and worms? With so many awkward questions to address it is easy to imagine why we avoided discussing the topic in the past. But evading the question is hardly the way to move forward. If we are ever going to find a good, or at least a better way of assessing where the line should fall, we need to be working on the problem, not ignoring it.

In choosing to tackle the fish pain question, however, we must acknowledge the sensitivity that surrounds investigations of this nature. To determine whether an animal feels pain, we need to find ways to induce something we agree is pain. However, it is ethically and morally challenging to design experiments whose very purpose is to cause harm. Yet we must, if we are to learn whether an animal has a capacity to suffer and so whether we should protect it. Scientific research in this area is both strictly regulated and closely scrutinized. Considerable efforts are made to protect vertebrate animals used in this way. Before any experiments can begin special permission must be granted from a number of different bodies and permits and licenses must be obtained. The way this is done varies from country to country, but what each has in common is the aim of limiting the potential suffering that an animal is exposed to.

In Britain, the 1986 Animal Scientific Procedures Act was passed to regulate how animals are used in experiments, and it is very specific about what is acceptable practice—the Act aims to minimize pain, suffering, distress, or lasting harm. Long before experiments can begin, researchers must complete several days of training and then pass exams, including a hands-on practical test, to ensure that they are aware of the legislation and that they know about the biology of the animals they will work with. On passing the exams the researcher obtains a licence to undertake animal research, but before becoming fully independent they still need to complete a probationary period with an experienced animal handler overseeing their work until the researcher demonstrates a sufficient level of competence. Those who manage research programmes must take additional training in ethics, experimental design, and statistics.

Before a research project begins, it is their responsibility to write a proposal in which they carefully justify the questions they want to address and the methods that will be used. Part of this justification requires that alternative solutions be considered and that the scientific gains be weighed against the suffering incurred.

These project proposals are thoroughly screened by ethical review panels. Such panels are made up of scientists, administrators, lay-members of the public, and governmental representatives. They consider the number of animals that will be tested and the methodologies and protocols to be used. These are then validated against the potential benefits that the results may provide. Many scientists complain about this lengthy process, but the regulations are important—the training and the writing of the project proposal force researchers to contemplate the real value of the animal work being proposed. Sometimes the review process deems that the work is appropriate even if it means a number of animals will experience pain. The suffering of a few animals within the context of the current experiment is justified on the basis that it will help to treat or alleviate future pain and suffering for others, usually humans but sometimes other animals too. However, there are also cases that, upon reflection, are considered unacceptable, and in these situations permission and permits for the research are denied.

In Chapter 3 I describe the work that my colleagues and I did to determine whether fish feel pain. Prior to starting our research we had to obtain permission from the UK Government. All aspects of the work were carefully and conservatively designed. The ethical issues were openly discussed as part of the application process for the permits. And throughout the work, my colleagues and I made sure that we minimized the numbers of fish used and we strove to use pain stimuli that would be mild to moderate. Our research took a cautious approach that in essence boils down to three separate questions. These built on each other in such a way that it only made sense to proceed to the next question if the answer to the last one was found to be true. We began by simply asking, do fish have the necessary receptors and nerve fibres to detect painful events? Next we wanted to determine whether a potentially painful stimulus triggered activity in the nervous system. If we were able to find positive answers to those two questions, the final test was to find out how the experience of a potentially painful event affected the behaviour of fish and the decisions that they made.

Using these different steps we incrementally built up a picture of how fish detect and respond to something that damages them. The first two questions were fairly straightforward requiring 'yes' or 'no' answers and the tests did not require live, active fish—they used tissue samples or fish that were deeply anaesthetized and would never recover. The final question, however, was harder to tackle and the results were the most difficult to interpret. The last phase of the work addressed whether fish show signs of suffering. As we will see in Chapter 4, this is a challenging question because to show that fish suffer we need to ask whether they are sentient—do they experience feelings and emotions, and if they do, does that mean they are also conscious? Can we ever really know what another animal actually experiences? This is a question philosophers have pondered a great deal and it turns out to be central to the fish pain debate.

In the mid-1970s an essay written by the philosopher Thomas Nagel asked whether it was possible for us to ever truly know what it would be like to be a bat. Nagel used this idea to emphasize how consciousness is a subjective state. He warned against trying to reduce the inner experiences conferred by consciousness into objective terms. He used the example of the bat to illustrate the gaps in our understanding of the philosophy of mind. Although bats are warm-blooded mammals, they are very different from us: they fly and employ ultrasound to help them navigate and capture prey, and so they have skills beyond our own subjective experiences. Nagel did not deny the bat its own experiences or subjectivity—he simply stressed that we will never experience a bat's subjectivity for ourselves. Nagel's ideas are still debated today, and what we mean by consciousness remains unresolved. Nevertheless Nagel's opinions are useful for the fish pain debate—we may never have the opportunity to fully recognize what fish experience, but we should not deny them a capacity for subjective feelings just because we cannot experience their feelings ourselves. In many contexts we don't. Consider the permits and training my colleagues and I needed to obtain for our research—the 1986 Animal Scientific Procedures Act protects *all* vertebrates. So, even before there was any evidence for or against pain perception in fish, they had been recognized as a group that should be treated in ways that minimize their potential pain and suffering.

Given that fish are already legally recognized as requiring protection with regard to animal experiments, it is all the more curious that our 2003 article on pain in fish attracted so much attention from the world's media. Why did headlines reporting fish feel pain sell newspapers? It seems to come down to the fact that people consider fish to be different; they're... well, they're fish. They fascinate us, but there is something curious and perhaps a little unsettling about the topsy-turvy way that they exist. They are inextricably tied to the water and literally suffocate in air, and while they have a face with eyes, nostrils, and a mouth, these features appear to be rigid and fixed, which contrasts sharply with the more expressive, mobile faces of most terrestrial vertebrates. Fish also have a number of alien senses that they use to detect the world around them.

We have five senses: hearing, smell, taste, touch, and sight, which we rely on heavily to guide us through the environment. Fish have all our senses but they also have more. When you view a fish sideways on, for instance, you can make out a thin line that runs along its flank from just behind the gills towards the tail. This is the lateral line—a pit filled with special sensory receptors some of which allow the fish to detect nearby objects. It's a way of 'seeing' without eyes. One group of fish, aptly called blind cave fish, live in underground caverns in Mexico where it is so dark that eyes are useless and the fish have quite literally lost them. Yet when you watch small groups of these curious-looking fish swimming around a tank it is quite obvious that they know exactly where they are in relation to the walls, to other fish, and to the various objects within the tank. The fish don't collide with things because as they swim they interrogate the area around them using their lateral lines. As the fish swim forward they set up a bow wave in front of them—just like a boat does. As this bow wave interacts with solid objects close to the fish, parts of the wave are reflected back to the lateral line where special sensory receptors detect the wave patterns. The cave fish are able to translate the reflected waves into information about the objects around them. Three specialized nerves convey information from the lateral line to the brain where there are areas specifically devoted to processing this information. Thus these blind fish can readily build up an internal image or map of what or who is close by. Fish species that have functioning eyes also have a lateral line, and though not quite so dependent on it as the blind cave fish, their lateral line provides them too with this additional sense.

Other fish have developed ways of both generating and sensing electricity. A specialized electric organ located towards the end of the tail can, in the case of the electric eel, generate sufficient electricity to stun prey. But some species, such as the knife fish or elephant nose fish, generate weaker electrical signals that they use for communication: the frequency of electrical impulses acting as unique identifiers for different individuals. These fish also use specialized receptors embedded in their skin around the head to pick up weak local electric fields created by prey animals. This electricity-based sense permits the fish to hunt for prey in the murky waters of the Amazon where, like the lightless caverns in Mexico, eyes are next to useless.

These curious sensory systems are so different to anything we possess they emphasize how different fish seem to be, but if we take a closer look, are fish really all that different? Apart from the obvious backbone, fish have plenty of characteristics in common with other vertebrates. Their overall physiology, for instance, shares similarities with processes seen in other vertebrates—even us. The way that they respond to stressful situations, the so-called 'stress response', is strikingly similar to the way mammals cope with stressors. After experiencing a stressful event our bodies release more cortisol into the blood, and the same is true in fish. To manage the stress response and assist the body's return to normal we have various feedback mechanisms that help to control our response—and fish make use of very similar processes.

The fish brain also displays attributes similar to those found in other vertebrates; the main divisions we recognize in ourselves are there in fish—a forebrain, a midbrain, and a hindbrain. More specifically, the functions of certain structures are also remarkably similar—in Chapter 4 we will delve deeper into the evidence for this and explore how evolutionarily conserved some brain structures turn out to be. Yet despite these vertebrate commonalities, you cannot escape the fact that when you see a fish brain it looks strangely different—to be blunt, it looks naked. This is because fish do not have a neocortex, the grey matter that gives our brains its characteristic, crinkly, convoluted appearance. This missing structure plays a central role in the fish pain debate—those who argue fish cannot feel pain consider the neocortex to be essential for an animal to experience feelings. Certainly magnetic resonance imaging (MRI) studies, which allow us to observe areas of brain activity in real time, show our own neocortex to be active during painful events, but the activity isn't exclusive to the neocortex; other parts of the brain are also working. Various brain imaging techniques have revealed that these areas lie beneath the neocortex—and some activity occurs in structures also found in fish brains.

Moreover, arguing that a missing brain component—such as the neocortex—prevents animals from performing certain kinds of skill or activity may not be a productive line of reasoning. As brains have evolved and become more complex, 'newer' areas achieve some of the functions previously performed in older areas. But we need to be careful in how we compare brain function and brain capacity across different taxonomic groups. Rather than guessing what different brain structures may or may not confer, a better approach is to formally test the function of different parts of the brain and then determine how these influence the behaviour or the choices that animals make. It is only recently that this kind of methodology has been applied to fish.

Just a single example shows how misleading it can be to infer that something processed by the neocortex cannot be achieved by a simpler brain. Brain laterality describes how we process different types of information on the two sides of our brain. We know from stroke victims who have suffered partial brain damage and also from studies that measure brain activity, that language and related information are typically processed on the left side of our brain, whereas thinking about visual information or recognizing faces are tasks that we usually process on the right. The two halves of the brain function differently and these differences in activity are happening within the neocortex. Using the same logic that the absence of a neocortex makes it impossible for fish to feel pain, we could argue that fish will be incapable of brain laterality as well. But it turns out this is wrong. Fish lateralize different categories of information in the two sides of their brain.

With numerous colleagues, Giorgio Vallortigara from the University of Trieste and Angelo Bisazza from the University of Padova, Italy, have discovered that several fish species lateralize visual information. Some fish prefer to look at their shoal mates or neighbours using their left eye, and use their right eye to look at things that make them wary—such as predators or novel objects. Dividing information processing between the two halves of a brain increases efficiency because each side of the brain can simultaneously work on different types of information. This is literally parallel processing. Coping with at least two pieces of information at a time might be essential if you live in a dangerous environment where hungry mouths lurk in the shadows. It is an important ability—and you don't need a neocortex to do it.

Fish might share similar physiological mechanisms with other vertebrates and exhibit brain laterality like our own, but conventional wisdom portrays them as foolish creatures with a three-second memory. This view is wrong. In Chapter 4, I develop the concept of fish cognition and what fish can achieve. Examples of intelligence in fish serve a number of purposes. Demonstrating that fish are clever in ways that resonate with our own behaviour helps make them seem less alien. Furthermore, in searching for evidence that fish are sentient we need to demonstrate that fish have an ability for complex cognition. We need to show that they can integrate different sources of information to help them evaluate situations and then respond appropriately. Several fish species are surprisingly smart and research has shown that they have accurate memories that can last several days, or even years in the case of migrating salmon. Juvenile salmon learn the sequence of smells and odours that they encounter on their first migration out to sea. On their return as mature adults, which can be one or more years later, they recall the list of smells that they memorized and use this to help them navigate back to the very same stream that they hatched and grew up in.

Today, we interact with fish in multiple ways. Many of us keep them as pets—a pair of goldfish in a bowl or perhaps a tank of brightly coloured tropical fish. These can certainly be an eye-catching addition to a living room, but our desire for such ornaments can be costly. Certain fishing practices used to capture wild tropical reef fish, for instance, are taking a serious toll on the world's coral reefs. In South East Asia illegal fishing with cyanide strips reefs of their inhabitants and kills the corals too. The majority of the fish caught this way survive only long enough to be transported to pet stores where they are sold to customers, but many die quickly afterwards from slow cyanide poisoning. Several conservation groups are currently focusing on alternative ways of capturing or rearing fish for the pet trade, but for many reefs it is too late.

Increasingly fish are also used for research—in many cases they are now the preferred animal for biomedical and chemical testing. Not that long ago we used mice for this task, but a commitment to reduce the number of rodents in laboratories has meant that much of the routine biomedical and toxicology screening is now done with a small striped, silvery-blue fish called the zebrafish. This transition has been so popular that many research labs have swapped their racks of mouse cages for shelves of small aquaria. The discovery that several human and zebrafish genes share similar kinds of function has allowed this model species to be used in studies of human diseases such as cancer. Once again we see that fish are not so different after all.

The past decade has seen a general recognition that there is a need to find alternative solutions to the use of rodents in research. This has been driven by an increased awareness and respect for animal welfare and the recognition that scientists should work towards achieving the 'three Rs': replacement, reduction, and refinement. William Russell and Rex Burch synthesized this concept in their 1959 book, *The Principles of Humane Experimental Technique*. In this they urge researchers to search for methods that avoid the use of animals, but if animals cannot be replaced then it advocates the use of careful experimental design and powerful statistics to reduce the overall numbers of animals required. Further, it encourages scientists to improve experimental approaches so that the negative impact the research has on animal welfare is limited. Animal welfare concerns more than just health and well being; it expresses ideas about the quality of an animal's life and maintains the moral view that animals that are sentient should be protected from unnecessary pain and distress. In some sense, fish have a nervous system simpler than that of rats and mice, hence the label 'lower' vertebrate, but is the substitution of fish for rodents really 'replacement'?

Our commonest interaction with fish is that we fish for them—angling is one of the most popular leisure activities in the UK outside of the house, and it is a major sport and hobby supported by an enormous global retail industry. The attraction of this pastime comes through the challenge of the pursuit and the desire to outwit the fish using a skillfully handcrafted fly or a bait that the fish will be tempted to bite. Angling demands a high degree of patience and a good level of understanding of fish behaviour. A bewildering amount of fishing gear and gadgetry is now commercially available; for instance, attaching a small portable sonar device to a rod can help reveal where the fish are lurking. Often a goal of the angler is to catch fish to eat, but the process of 'catch and release' fishing is becoming increasingly common. Sometimes this is a conservation tool to help manage fish populations—fishermen are not allowed to take the fish they catch, so they unhook them and return them to the river or lake. Whether this actually is an effective conservation tool is contested and it is also not clear what the welfare implications of release really are. Many anglers who fish as a hobby fear the fish pain debate could generate political and legal challenges to a number of current practices. Animal welfare considerations have previously brought about change, such as the banning of bear baiting, cock fighting, and more recently in some countries, certain forms of fox and deer hunting.

Yet many of the same fishermen who fear the fish pain debate choose to use barbless hooks to help them quickly remove the hook from the fish's jaw, minimizing the handling time and the level of stress induced in the fish.

But, in terms of sheer numbers of fish, the real business is ocean-going trawlers scooping fish from the sea. Fish, netted by the tens of thousands, are pulled to the surface through such rapid changes in pressure that their swim bladders overinflate, causing the body to become severely distended. On reaching the surface, the fish are dropped onto open decks where they then flap around as they suffocate. We tend not to think too hard about the way we capture fish at sea—it isn't very pretty. We wouldn't accept killing chickens by throwing them into a tank of water and waiting for them to drown, so why don't we object to fish suffocating on trawler decks? Being pessimistic one might argue that the implications of the fish pain debate for commercial fishing might soon be irrelevant. As wild fish stocks come under yet greater pressure from over-fishing it becomes more and more expensive to harvest wild fish, so much so that increasingly we now farm fish—captively breeding and rearing domesticated strains for our consumption.

Aquaculture is the fastest growing form of farming across the globe. The number of fish farms and the number of species we farm continues to expand as we become more and more dependent on these methods for providing us with fish to eat and to supplement wild fisheries stocks.

Fish farming has roots going back several millennia; there are references to the use of sluices and fish ponds in the Bible (Isaiah 19.10), and over three thousand years ago the Chinese gathered fish after flooding events and transferred them to ponds where they were fed on a variety of foods including waste material from the silkworm industry. Production of fish through these early forms of aquaculture was seen as a way to reduce the effort of seeking out and capturing wild fish.

Modern aquaculture is an industrialized process where a small number of companies manage, on a global scale, large facilities that produce millions of tonnes of fish a year. The shift from the fisherman that heads out to sea to harvest wild fish to the fish farmer that maintains his stocks just off the coast in large, deep pens goes some way to explaining why we now talk about fish welfare. In the same way that terrestrial farming has faced enquiries from concerned consumers about farm animal welfare, aquaculture is beginning to experience some of the same scrutiny. But of course aquaculture is just one of many ways that we have an effect on fish, and as we ask questions about fish welfare within the context of a fish farm we open up a debate that is more wide-ranging. If we are concerned about the welfare of the fish we farm, then logically we should be concerned about the fish we catch on a rod and line, the ornamental pets that we keep in our living room, and the fish we use for chemical testing in the research laboratory.

Over the past half-century our attitudes and perceptions towards animals have changed. Natural history documentaries and scientists such as Niko Tinbergen and Jane Goodall have opened our eyes to the complex and fascinating aspects of animal behaviour. We have discovered that animals are more sophisticated than we have previously given them credit for, and in the case of Goodall's work with chimpanzees, we see that they can be startlingly like us in terms of their social behaviour and their individual personalities. As our knowledge has grown, our perception of animals has shifted. The general public now has a view on how animals should be treated—caring appropriately for the animals we interact with now appears to be a basic concept of humanity.

The end of the Second World War was a time of many changes, including a massive transition in farming practices. Small, individually owned farm operations were lost, so that instead of rearing a few chickens or a small herd of cattle, animals became mass produced in industrial systems. While the increased productivity led to cheaper food, some began to query how the animals coped in the new and very unnatural environments. Concerned consumers wanted to know about the produce they purchased. Confined housing may deliver the efficient production of meat and dairy goods, but what were these unnatural housing conditions doing to those animals? It was these kinds of questions that put animal welfare, particularly farm animal welfare, on the map.

In considering the welfare of animals that we interact with, we might want to know whether they have sufficient access to food and water, do they have somewhere suitable to rest or shelter, are the animals healthy and free from pain and injury, can they behave normally, and are they able to avoid fear and suffering? This is a summary of the 'Five Freedoms' put forward by the Farm Animal Welfare Council, an independent advisory board established by the UK Government in 1979. Some aspects are easier to resolve than others; providing access to food and water for example is straightforward. But determining whether the animals are able to behave normally is much harder—what if we are asking about an animal that naturally has a large home range? By keeping the animal captive we prevent it from traveling long distances—is this affecting its welfare?

The five freedoms are also hard to apply beyond farming. For instance, it has been proposed that fish housed in laboratories should be given tanks with 'enrichment'—i.e., their tanks should be furnished with gravel, plants, and other structures to break up the spatial environment. The enrichment is believed to be an important addition that will allow fish to behave normally. The problem with enrichment is that some species become aggressive when there are areas that can be defended as territories, and adding plants or pebbles into a tank gives these fish an area or an object to protect—something to fight over. So while the enrichment may be beneficial for some species, it could potentially make the environment worse for others. Normal, territorial behaviour in this instance doesn't mix well with the close confines of the captive environment.

Animal welfare is hard to define, because the way animals behave and respond to different kinds of environment or stimuli are species-specific and context-dependent. For long enough the science of animal welfare was thought of as woolly, and the failure to pinpoint what we mean by good welfare did not help its image. But things have improved and increasingly we now use well-designed scientific experiments to investigate welfare issues. These have allowed us to explore animal needs and preferences, and so to find out what animals want in their captive environment. Animal welfare science provides us with tools to determine what kinds of experience are good for the animal, and conversely what are bad. Finding out what triggers or contributes to animal suffering allows us to find ways of avoiding it. Over the past two decades, sound scientific practices have helped establish what good welfare means for terrestrial animals, and we have refined many practices to help relieve animal suffering.

As we try to establish what good welfare for fish might mean, we can look back at the kinds of experiment that helped determine good welfare for terrestrial animals. Using these to guide us we can start to ask what fish need, what they prefer, and what is detrimental to them. And I explore this in the final chapter of the book, but before we do

that we must determine whether 'fish welfare' is a meaningful term. Is it something we should strive for, or do we
need to go there at all—can we dismiss it as irrelevant? To
answer this we need to find out whether fish have a capacity
to suffer.

This book examines the evidence we currently have for
whether fish can suffer, and indeed whether it is meaningful to discuss pain in fish. My goal is to provide you
with sufficient information that will allow you to make up
your own mind. Pain and suffering are at one end of the
animal welfare spectrum, whereas we might place health
and well-being at the other end. Ethically we may feel
obliged to ensure any animal we care for is healthy, but
beyond that do we need to bother ourselves with questions about suffering and pain? A recent move within the
field of animal welfare has proposed that we should
simplify our approaches and merely ask, 'Is the animal
healthy?' and 'Does it have what it wants?' This approach
was introduced by Marian Dawkins, a research professor
in animal welfare at Oxford University. Her goal was to
simplify animal welfare assessment criteria.

While Dawkins' two questions parse the problem down
to something we can directly ask of cows in a dairy herd,
this approach is still limited in where it can get us. The
problem is that we do not think of all animals in the same
way. For example, we could ask these two questions for
beef cattle confined to stalls, chickens housed in battery
cages, fish kept in sea-cages, or shrimps maintained in

culture ponds. We might be willing to extend welfare considerations to the cattle and the chickens, and you will decide over the course of this book whether we should offer similar protection for fish and what it might mean if we did. But shrimps—do they warrant welfare considerations? How do we decide where to draw the line? To answer that we must decide what criteria matter, what aspects allow us to label an animal as requiring good welfare. There are several ways to tackle this, but more often than not the starting place is to ask, does the animal suffer from pain and injury? Animal welfare is now a large research discipline with dedicated scientific journals that publish research papers on a wide variety of topics. The science of animal welfare goes beyond pain—and this book will too. But we must start with pain and the effects it has on different animal groups.

What Is Pain and
Why Does It Hurt?

Pain is a personal experience; yet we are reasonably confident that other humans also suffer from this sensation. Through discussion, we can share with each other the feelings that it generates and we can learn how to treat it or get relief from it. Sometimes doctors or physiotherapists ask us to describe our pain and certain diagnoses require that we score the intensity with which we feel it. When faced with such questions we find it is possible to distinguish between different forms and level of pain and we use words such as dull or sharp, general or localized, and aching or throbbing to illustrate what we mean. But do animals experience the same kinds of sensation? If they don't, then perhaps there is no need to be concerned about a dog that is limping because it has a gash on its paws, or a battery caged chicken that has a broken leg. And what about fish—if they had a wound on their flank, would they experience that as painful?

The dog and the chicken might flinch and even yelp or squawk if you touch their injured limb and this often makes us feel uncomfortable. It seems that we naturally empathize with injured mammals and birds, and typically we interpret their situation to be one of pain. We can imagine how a damaged limb would feel to us and we recognize that we too would flinch if something or someone forced us to move. So does the fact that animals try to protect a damaged area mean that they feel the pain? How can we tell what goes on in an animal's mind?

The animal pain question is made even harder because we still don't fully understand how pain works in our own bodies. We have a pretty good idea of how the nervous system detects and responds to painful events, but how the brain processes the information and generates emotional responses associated with the hurting and suffering induced by injury is less clear. All of this uncertainty makes it difficult to work on animal pain. Scientists like to parse processes down to clear-cut events where we can state; 'Yes, it does occur,' or 'No, it does not.' Studying animal pain is awkward because the preferred, crisp, black-and-white scientific response gives way to a more blurred, 'Well, it might.'

One way to study whether non-human animals experience pain is to ask why it has evolved. It is unlikely that pain spontaneously arose during evolution only when humans appeared, and so we might expect to see similar kinds of process in at least some other animals. The pain an animal experiences could well be different to the pain we experience, but it seems unlikely that there will be a complete absence of pain-like processes in animals. Evolution typically works through small changes that become honed into effective adaptations by natural selection—gradual changes that lead to new structures and new ways of behaving. So searching for pain-like processes in other animals should allow us to find something, even though it might be much simpler than the processes we know and recognize in ourselves. Perhaps if we survey animals from across the animal kingdom, it will be possible to trace the evolutionary history of pain. Where did pain first arise, in mammals, in vertebrates, or does its history have even older roots than animals with backbones?

There are good reasons for thinking it will have very old roots. Pain helps us recognize when we need to change what we're doing. When we are injured, it lets us know that we need to rest or protect the injured area to prevent further damage. It could be described as a process favoured by natural selection—an animal that minimizes the harm it does to itself will heal more quickly and recover faster and so go on to leave more offspring than one that cannot. So being able to experience pain, while unpleasant in itself, is almost certainly advantageous in evolutionary terms. Thinking along these Darwinian lines, it would be very strange if pain or pain-like sensations were *not* found to be widespread phenomena across lots of different animals.

But even if we find pain-like responses in animals, do those animals actually suffer from the experience? If I wanted to explore your experiences of pain, I could ask you to describe something that had previously hurt you. Your explanation would allow me to assess how intensely you felt the pain, where it hurt, the time-course of the event, and even what form the pain took. We can use language to share our feelings with each other. Without the benefit of language, however, we need ingenious experiments designed to reveal what an animal is experiencing. We can begin to get an idea of this by watching changes in their behaviour or alterations to the decisions that they normally make.

In the simplest of terms we are trying to determine whether the animal is happy or miserable. Terms such as 'happy' or 'miserable' might be strictly inappropriate to use when interpreting the behaviour of animals, but sometimes it is easier to use these terms in a general way to indicate that the animal seems to be in a positive frame of mind because it is playful and interactive. Or alternatively, it seems to be in a negative state of mind because it is subdued and withdrawn. We do need to take care when we use labels normally associated with human emotions, but we can recognize when animals have positive and negative moods. Internal states and emotions are a key component that influence how we experience and feel the pain process.

What a number of researchers have shown over the past two decades is that while it is a challenge to get inside the mind of an animal, it is not impossible. With the right design of experiment and armed with conceptually simple questions that animals will understand, we can find ways to sidestep the language barrier. For instance, researchers have used experiments where there are two or more choices to determine what animals prefer—these trials reveal which option or item the animal selects. We can even use these kinds of approach to find out how strongly an animal desires certain options.

When you really want something you find yourself motivated to get it; this might mean that you are prepared to pay a bit more for it, or perhaps go further out of your way to find the right place to get it. Animals will similarly go out of their way or work a little harder to get access to resources that they value. So-called 'choice experiments' have proved to be very useful in animal welfare research. Demonstrating that an animal will work a little harder to have access to a companion, for instance, tells us that the companion is important to the animal being tested. Sometimes what we find is expected. Battery chickens, for example, really value the ability to dust-bathe; preening and cleaning by dust-bathing are something wild chickens do on a regular basis. Other findings, however, have been more surprising: farmed mink value baths of water in which they can swim more than access to pens that have toys for them to play with or tunnels to investigate.

Testing animal choices using these kinds of approaches can also to tell us what an animal would prefer to avoid. Rats, for example, can be given choices between two kinds of drinking water, one sweetened and the other flavoured with an unpalatable drug. Normal, healthy rats will avoid the unpalatable water and choose to drink the sweeter tasting option. This preference can be changed though. If rats with arthritic joints are given the same options, they select the water that contains the pain-relieving drug even though it tastes bad. The fact rats choose to self-administer the pain-relieving drug appears to be compelling evidence that rats with arthritis do experience a sensation akin to pain.

To get a clearer understanding of how pain sensations are generated, let's take a very simple scenario and describe how pain arises in us. Imagine that for the briefest of moments you forgot that the pan of water you are bringing to the boil has an old, poorly insulated lid. To check whether the water is boiling yet, you pick up the metal lid and look into the pan—there might be a slight delay but very quickly you will drop the lid before you even think about it. At this point the burn you have just inflicted on yourself has been unconsciously detected; tiny receptors on the skin of your hand have been triggered and these generate electrical impulses that pass through single fibres within nerves that connect to the spinal cord. A reflex response is automatically put in action, and a signal quickly passes to the muscles that control your hand to tell you to let go of the pan lid. Now you might begin to experience or feel the pain. You become aware of the unpleasant, burning, throbbing sensation in your hand—it hurts. It may cause you to cry out, and almost certainly you'll pull your hand towards you and survey the damage. There may not be too much to see, some reddened skin, a bit of swelling, but the pain you will start to feel will be intense. You might look for something cold to press against the swelling, or you could put your whole hand under running, cold water to try and relieve the pain. Certainly the next time you need to lift the lid on that old pan you will remember the burn and you will be reminded to use a cloth on the handle to protect your hand—it was a painful lesson and it will stay with you for quite a while.

Splitting this scenario up into its different parts we see that pain is not just a single process, but rather a series of separate events. The special receptors in the skin are stimulated, telling the body something is damaging it. The burn is initially detected and a signal conveying this information travels to the dorsal horn of the spinal cord where a reflex response is triggered automatically, making you drop the lid—so far all of this occurs unconsciously. You don't think about it, you just do it. At this point, you have experienced nothing we would recognize as suffering. After the signal has reached the spinal cord and the reflex response has been triggered, the signal moves up to the brain. Only after it reaches the brain do we begin to *feel* the pain. Now we become consciously aware of the emotional, unpleasant sensation associated with the burn—the brain is telling you that whatever you just did, hurt. Depending on the type of tissue damage it can take as long as two seconds for some forms of pain to be felt, but for most injuries it is faster than this.

The first, unconscious phase to this process is known as nociception—'noci' relates to injury or damage and 'ception' refers to perception or detection, so it literally means detection of injury or damage. In mammals and birds, as nociceptors on the skin are stimulated, electrical impulses begin firing in specific nerve fibres dedicated to transmitting information about tissue damage. Once the signal reaches the spinal cord a reflex response occurs. The process of nociception is relatively straightforward in that researchers working on pain have discovered how the sensory receptors are triggered and how they respond to different types of damage. It is the next set of processes, the movement of the signal up to the brain and the conscious detection of pain—the part that allows us to recognize something is hurting and by how much—that is, by contrast, more of a black box.

When we talk about a pain we feel, we portray sensations and emotions that have been generated by our brains. So the pain we describe to one another and the feelings that we empathize with are all part of the conscious phase—the bit that happens beyond nociception. As we'll see shortly, nociception is a key process for many animals, not just humans. But just because an animal detects injuries through nociception does not mean that they feel pain too. The key question is whether they, like us, have a conscious experience that the damaged area hurts. This question lies at the heart of the problem. What we are really asking here is 'Are animals aware that they are sore and in pain?' Or put another way, we are asking, 'Are animals consciously aware of the pain?' Whether sentience and consciousness are processes that occur in non-human animals is something that has occupied philosophers and psychologists for decades, and they have yet to agree an answer. I do not have one either, but this does not make the problem hopeless—it is possible to look for signs of awareness in animals. It is possible to design experiments that explore this—recall the arthritic rats that chose to self-administer pain-relieving drugs.

Before we start to search for examples of animal awareness, we need to consider further the role of nociception. To explore what the detection phase of tissue damage can do, let's look at which animal groups have nociceptive responses and see how they respond to damage. There are many examples of invertebrates that have nociceptive-like responses; from insects to worms. Nociception is a very common process, but given the fundamental protective role that it plays, that is hardly surprising. Animals with a nociceptive ability will have an evolutionary advantage over those that cannot detect damage. An animal that can recognize when something bad is occurring, and that responds by moving away from whatever is causing the injury, will have a better chance of surviving another day. It isn't hard to see why this is a process that natural selection will have favoured.

How far back in evolutionary history do we need to travel before we see the first signs of nociceptive-like responses? The answer is a surprisingly long way-simple nociceptive-like reactions can be seen in some of the earliest organisms with a nervous system. For example, we see them in Cnidarians, the animal group that contains jellyfish, corals, and sea anemones. These apparently simple invertebrates possess diffuse nerve nets that allow electrical signals to pass through their bodies. The nerve nets consist of interconnected neurons, but there is no brain and no specialized areas with clusters or bunches of neurons. Despite this very basic setup, the nerve nets can trigger responses such as body contractions that allow these creatures to move away from danger. Toxic chemicals or attack by a predator leads to evasive responses very reminiscent of nociceptive processes.

You may have even experienced all this at first hand; if you've ever tried to prize a sea anemone away from the wall of a rock pool you will have witnessed it first withdraw its tentacles and then shrug away from you. This is the way it protects itself; it detects that damage is occurring and so tucks away its vulnerable, soft tentacles within its more robust body wall. In fact, it turns out that these mechanisms are not only there to protect the anemone from would-be predators, but also help to protect anemones from each other. These sedate looking animals have a Jekyll and Hyde existence—while they may appear to be peaceful creatures with tentacles that gently waft in tidal flows, the truth is that they can be incredibly aggressive with one another and sometimes even fight to the death. Contests usually arise when a neighbouring anemone has got too close for comfort. Fights can even be set up, allowing us to study them.

If you carefully collect a few anemones from a rock pool and place them into a bowl of cool sea water with small pieces of slate or stone, they will select a piece and attach themselves to it. They have a muscular foot that allows them to creep around their environment if they need to, but it also helps them to firmly attach and anchor themselves when they find a suitable spot. Once the anemones in the bowl have selected a stone, try placing two similar sized animals next to each other and watch the battle commence. After detecting each other's presence and perceiving themselves to be too close to one another, they will start to reach out with their tentacles and begin grappling. During these interactions some species have a special set of fighting tentacles that now appear. These are normally hidden, tucked away under the regular food-gathering tentacles around the collar at the top of the column that forms the outer body wall. The fighting tentacles, known as acrorhagi, are filled with special stinging cells called nematocysts that the anemone uses to attack its rival.

Fights can be prolonged, but ultimately they end up with one anemone as the victor while the loser slowly edges away. Even simple animals such as sea anemones recognize when they are under attack and, if the attack is overwhelming, they withdraw. Is this response nociception? It is different to the nociceptive response that we have, because the simple nerve nets of the anemone are not relaying their signal to a specialized location in the nervous system via neurons that coordinate a response. The anemone simply reacts by pulling its delicate body parts in to protect itself. This less specific series of events is what I have called 'nociceptive-like'. Considering how an anemone reacts, it does fit the general definition of detection of damage or injury by an animal's nervous system. It's also worth noting that while the pair of dueling anemones wrestled with each other they made use of hundreds of stinging cells on the acrorhagi tentacles. The stinging cells work by firing a structure containing a toxin into their opponent. These are very similar to the stinging cells found in jellyfish tentacles. We don't know whether the anemones detect the stinging itself, but the toxins released are neurotoxins designed to impair nervous systems—like those that hurt us.

Continuing the search for nociceptive-like responses, scientists have investigated whether there are any similarities with the neural mechanisms associated with pain processes in ourselves. For instance, morphine is a well-known general pain reliever in humans. It works by binding to a specific set of receptors involved in nociception—the opioid receptors. When the morphine binds to the receptors, the pain signal has trouble generating a negative pain sensation. Our bodies use this system to help regulate our own pain—chemicals that we produce and secrete can block opioid receptors and this provides us with a way of naturally dampening pain sensations. The presence of opioid receptors in animals might also point to their having nociceptive abilities. Opioid receptors are certainly found in all vertebrates and, it turns out, in quite a number of invertebrates as well, including snails.

Snails have temperature-sensitive receptors in their feet that help them to avoid hot places. This can be demonstrated by putting a snail onto a metal plate warmed to about 40 °C. Within seconds of the plate heating up the snail responds by lifting part of its foot into the air. This is very reminiscent of a standard test used to study temperature sensitivity in rats and mice: a paw is placed onto a metal plate that can be heated to a specific temperature, the point at which the animal withdraws its foot reveals the heat sensitivity of the paw. It has been shown that snails change their response to the warmed plate if they are treated with chemicals that either enhance or prevent opioid receptors from working. We know that when mice are given a morphine-like chemical, their foot withdrawal is slower, but if they are given naloxone, a chemical that works in the opposite way to morphine, their foot withdrawal happens much more quickly. When the same type of study is done with snails, they behave just like the mice. This indicates that the snail withdrawal response involves pathways affected by opioid receptors. The nociceptive-like reactions in sea anemones and the presence of opioid receptors in snails suggest that nociceptive-like systems are very ancient in evolutionary terms.

Injury also triggers various protective responses directed towards the damaged area. Depending on which part of the body is hurt we might see changes in body posture, or the animal may try to cradle or shield the affected area, just like we do. Although somewhat anecdotal, grooming or rubbing areas of the body is believed to indicate some form of awareness of pain in animals, again, just like it does in humans. Plenty of examples of this can be seen in a broad range of animal groups and some forms of protection are more extreme than others. Very recent work has shown that even translucent, glass prawns commonly found in rock pools are aware of damage to their sensitive antennae. These delicate creatures show increased grooming of their antennae if these fragile sensory structures have been pinched or squeezed with tweezers. Their grooming, however, is much less intense if the prawns are treated with a local anaesthetic. We will explore this further in Chapter 5.

As nervous systems become increasingly complex we find more sophisticated forms of avoidance of injury. Spiders, for example, choose to lose a leg if it becomes badly injured—a process called autotomy. For spiders, wrestling insect prey into submission can be a tricky business, especially if the insect has a sting and venom like a bee. When bee venom gets into the leg of a spider, the spider responds by self-amputating its leg at the joint closest to where the venom was inserted by the sting. In fact, self-amputation can be provoked in spiders simply by injecting the same components of bee venom that cause us to feel pain when we are stung. Self-amputation is also found in other invertebrates such as crabs. While these responses may appear extreme, they allow the animal to either get away or prevent toxins in the venom from spreading to other parts of the body. Broadly speaking, autotomy is rather like the process of nociception, because the detection of the venom or injury and the subsequent loss of a limb allows the animal to avoid further damage.

In addition to nociceptive responses, our own bodies change in other ways when we experience something painful. Tissue damage or trauma, for example, can trigger various physiological changes: breathing rate may become more rapid, stress hormones are secreted, there might be a general loss of appetite, or we may even begin to feel nauseous. Many of these changes are associated with the body trying to cope with the injury. While they may not necessarily require conscious awareness, they are processes occurring beyond the reflexive nociception stage. There is ample evidence that mammals, birds, and even fish respond to injuries with elevated heart and breathing rate as well as increased stress hormone production; levels of adrenalin can quickly increase as can the levels of cortisol or corticosterone. Some of these physiological changes are presumably associated with pain-coping processes. Changes to an animal's physiology can influence certain psychological processes. For instance, an injured animal can have a heightened sense of fear. Pain-related changes in an animal's awareness are something we will return to in the next chapter.

The breadth of animals that can respond to damage or injury with nociceptive-like responses and changes to their physiology is quite striking. Even among most invertebrates we find protective mechanisms in place. But where does this leave us—have we found evidence that snails feel pain? No. Recall the distinction between nociception and pain: nociception is the unconscious recognition by the nervous system that damage is occurring somewhere, but pain is the emotional sensation that whatever is damaged is hurting. Determining which animals are capable of pain perception is more difficult because this is the part that, at least in humans, we believe requires consciousness. But, as I mentioned earlier, consciousness and its presence in non-human animals is hotly debated.

This is a key aspect to resolve because if animals are not conscious, then even though they have nociceptive-like responses and their physiology changes after they experience tissue damage, it is meaningless to consider that they *feel* pain. To answer the question posed by the title of this book, we need to show that fish have the apparatus to detect noxious stimuli—the nociceptors and fibres that conduct the information—but we also need to ask whether fish are capable of perception and awareness.

There are some intriguing similarities between this problem and another similarly perplexing issue—that of whether unborn or premature babies feel pain. Certain surgical procedures that need to be performed on preterm babies or babies in the womb are of concern because when these are carried out on older babies and children we consider them to be painful. Surely then, we should find ways of delivering pain relief to the babies before the procedures are done. Many people would intuitively argue that we are morally obliged to protect premature or unborn babies and provide them with pain relief, but astonishingly, until very recently, there wasn't a medical consensus on this. There were some clinics and hospitals where pain relief would automatically be used, but others where it would be withheld. Seeking evidence to determine the right course of action, pediatricians have focused on two questions. First, do signals generated by nociceptors in premature babies reach the neocortex of the brain, the area in humans crucial for detecting sensations? And second, to what extent do we consider premature and unborn babies to be conscious? Earlier in this chapter we discussed the evolutionary advantage that detection of damage and even pain perception can provide to certain animals—but in the case of unborn babies, it's hard to imagine a selective gain for neonates to feel pain.

Researchers explored the premature baby problem in several different ways. The first question was investigated using a brain imaging technique similar to magnetic resonance imaging (MRI) but a little less accurate in how precisely it detects neuronal activity. This approach clearly showed that a heel lance, a cut made to the heel to draw blood for various tests, does generate a signal that can be seen in the part of a premature baby's neocortex that processes bodily sensations—the somatosensory cortex. By contrast, however, firmly gripping the heel did not produce the same brain activity, even though babies typically pull their feet away and try to withdraw from the hand gripping the heel. Observations like this have had several effects. They have fuelled an interest in developing drugs that will be safe painkillers for premature babies, and they have prompted further studies of responses to procedures such as heel lancing when painkillers are provided. But most importantly for current purposes, they have given health care workers in many premature baby units sufficient evidence to believe that these preterm babies deserve the benefit of the doubt. Most specialists now give pain relief to premature babies when possible. Yet none of these approaches or conclusions addressed the second question—which I suggested is the key one for animals—are premature or unborn babies conscious? It isn't clear how this could be tested—but activity in the somatosensory cortex was considered sufficient evidence for us to give human neonates the benefit of the doubt.

Could we apply to animals the same empirical approach used to reach conclusions in the case of premature babies? Is it possible, for example, to look at brain activity in different animals during potentially painful events? Imaging the brain during painful experiences has been done in a few animals, but it is a technique that has yet to be fully explored. This is partly associated with the cost, partly the logistics of keeping the animal still without requiring drugs to sedate or even anaesthetize the animals—hard enough in terrestrial animals. Trying to scan the brains of aquatic animals presents a whole new array of problems. Researchers are working to overcome these, but it's a tricky business.

If we can't view images of the brain, can we perhaps go back to older, more basic techniques used in earlier studies of neurobiology—can recordings of activity in specific areas be made with electrodes? Or, alternatively, is it possible to cut away or impair certain brain areas and then see how the behaviour of the animal is affected? For these methods to work, the brain must be organized in a way that we can readily find areas that we believe have a role in processing pain signals. Again, here we come up against some challenges in that as we move from our own complex brains to the anatomically different brains of animals, we need to be very clear about where we should look for activity. This becomes almost impossible as we move from comparisons of vertebrates to invertebrates, because invertebrate brains and nervous systems are so very different. If we keep the comparison within animals that have a backbone then the task is just about possible because all vertebrates have brains that can be divided into recognizable, distinct regions. We can make predictions about which areas of the brain we would expect to process pain-related signals.

At the beginning of this chapter I set out to describe what we mean by pain in ourselves—what mechanisms are involved and how it generates changes to our emotional state that leads to the pain hurting us and causing us to suffer. Pain is more than a single process. There is an unconscious phase where the nervous system automatically responds to whatever is damaging or hurting us; we call this nociception. Then there is a conscious phase where our brains become aware of the pain and we suffer from it. When we looked for evidence of nociceptors in animals we got quite a long way; even simple animals such as jellyfish and sea anemones respond in ways that suggest they have a basic form of unconscious perception and response to damage. The more difficult issue is to try and infer when, or where, in the animal kingdom nociception became linked to the emotional sensation of pain.

To find the answer to this we need to look for animals that have mental awareness. But even that isn't necessarily the whole answer, because studies in premature human babies found activities and responses in their brains that suggest they can process information relating to pain, but this is happening in a brain that is a long way from what we normally think of as conscious. So pain continues to generate more questions than answers. But let's have a go with fish. Let's look at what is known about their nociceptive abilities, and whether they can process emotions associated with pain.

Bee Stings and Vinegar:
The Evidence That Fish Feel Pain

In the last chapter, we learned that humans have specialized receptors and nerve fibres that respond specifically to injury or damage. Once triggered, a signal passes first to the spinal cord and then to the brain—only after it reaches the brain do we begin to experience pain itself. Armed with this knowledge of how pain arises in our own bodies, can we find out whether similar processes occur in fish?

In fact, this was the very question my colleague, Mike Gentle, and I asked each other just over a decade ago. Mike was based in the government-funded Roslin Research Institute near Edinburgh in Scotland—the same institute that had famously cloned Dolly the sheep a few years earlier. I was then at Edinburgh University where I was working on fish cognition and behaviour. Mike's research at that point was addressing pain in birds. He was investigating how pain in the leg joints of farmed chickens affected the bird's welfare. Our early conversations were mostly about our surprise at how little was known about pain in fish. It wasn't long before we realized that between us we could design a programme of work that would formally ask whether fish actually feel pain.

It seemed to us that the time was right to try to address the fish pain question. Globally, aquaculture was growing rapidly, but there was little guidance on how best to maintain, handle, and care for farmed fish. In contrast, the animal welfare movement had been influencing land-based farming practices for many years. Interactions with terrestrial farm animals have changed considerably over the past few decades, particularly as ways of reducing stress under intensive farm conditions have been found. To figure out whether welfare should be a consideration for fish, and particularly for farmed fish, the most obvious starting point was to ask whether, like terrestrial vertebrates, fish might suffer from pain. As it turned out, Mike and I were not the only scientists thinking along these lines. We didn't know it at this point, but a group in Russia was asking very similar questions.

Not long after our first meeting, Mike and I began writing a grant proposal that would allow us to investigate pain in fish. In the last chapter, I described how widespread nociceptors, or at least nociceptive-like processes, are in the animal kingdom; these are the first step in pain detection, so we fully expected fish to possess them. We were amazed to discover that there were no comprehensive reports on nociceptors or the specialized fibres that convey the nociceptive signal in fish. Could this be right? Was it really the case that at the end of the twentieth century we couldn't answer a straightforward question about whether fish had the necessary gross anatomy to detect pain? Fish are the largest vertebrate group. Did we really know so little? Ironically, we found more information on nociception in invertebrates than for fish. Over the following weeks we hunted down a number of old journal articles and reports, and as we dug deeper we discovered that a few scientists had tried looking at sharks and rays—fish with soft, cartilaginous skeletons—for the specialized nerve fibres that convey nociceptive signals. But there was virtually nothing about these fibres or the special forms of receptor in the far more common teleosts—fish with bony skeletons, like salmon, goldfish, and cod.

The studies that had found possible signs of nociception in some shark and ray species eventually concluded that the receptors were not responding in the way true nociceptors do. Receptors in the sting ray, for instance, did not respond to pressure stimuli—such as squeezing or pinching the skin—in the same way that mammalian receptors do, and they had no apparent reaction to temperature stimuli. These differences led researchers to conclude that sharks and rays do not have true nociceptors, at least not in the same way that birds and mammals do. When it came to bony fish, all we could find were some brief descriptive notes reporting how a number of free nerve endings had been observed in the skin of minnows, sticklebacks, sand gobies, and gurnards. Mary Whitear, a London-based researcher, had made these observations in the 1970s. She indicated in her report that these structures could potentially be connected with nociception, but she never went on to formally test this idea. Other than those general descriptions, we drew a blank. It really seemed to be the case that nobody had systematically asked whether fish can detect damage or injury through nociception.

We designed our research application around three goals. First, we wanted to find out whether bony fish possessed the sort of receptors and nerve fibres that control nociception in mammals and birds. Then, if these were there, we wanted to show that they were active when something damaged or injured the fish—we needed to find a way of measuring their response to different types of tissue damage. Our final goal was to move beyond the detection of the damage, so beyond nociception, to determine whether fish behaviour was affected by the experience of pain. Of the three parts, this last was going to be the hardest. The first two are relatively easy to resolve; either the characteristic nociceptor receptors and associated fibres would be there or they wouldn't, and if we found them, either they would respond to being stimulated or they wouldn't. Measuring changes in behaviour, however, was going to require more subtle measures.

To show that fish respond to painful stimuli, we would need to see changes in behaviour different to those which nociception alone might trigger. Our experiments would have to demonstrate that by applying a noxious stimulus to induce a potential pain response we could set off an initial nociceptive signal that might drive some form of reflex response. After that phase, and after the signal would have reached the brain, we would have to find a way of measuring how the fish altered their behaviour because of the pain or discomfort they experienced.

We decided that we needed to investigate two aspects of fish behaviour. The first explored behaviours closely linked to physiology. The behaviours that we selected were relatively simple, but, as they happened some time after nociception, they were going to be more than reflex responses. For example, we wanted to assess how a potentially painful event affected breathing rate and hunger levels. We selected these different aspects because we knew that they change in humans after something painful is experienced. While these responses might not necessarily require cognitive awareness of the pain, they can indicate the time-course of a reaction to the noxious stimulus. To try and pin down the cognition aspect, we devised a second approach to look for changes in so-called 'higher order' behaviour patterns. Here we wanted to explore aspects such as attention—were the fish distracted by a noxious stimulus? Humans find it difficult to focus and work through the pain of a headache—could Mike and I demonstrate something similar to this lack of focus in fish? If an experiment revealed that fish attention was diverted by experiencing a noxious event, we believed this would be good evidence that fish were perceiving the pain.

The next task was to decide which species to work with. After considering a number of different options we finally settled on trout. To us, they seemed a good choice because they grow to a reasonable size, allowing us to work with large fish. Their larger size would make finding and isolating individual receptors and nerves a little easier. The obvious alternative species were all much smaller, for example, goldfish or even the tiny but very popular scientific model, the zebra fish. Two other aspects attracted us to trout. In the UK they are a commercially important farmed species and they are also closely related to salmon. Salmon are by far the most popular, global species reared in aquaculture and, because they are closely related, we believed that what we found for trout would most likely apply to salmon. As one of our motivations to do this work in the first place was to learn more about the welfare of farmed fish, trout seemed to fit the bill very well.

The panel of scientists judging the merit of our grant application returned the first application to us, suggesting that we apply to them again, but focus the research on the mouth of the fish. In our original application we had emphasized the need to investigate damage close to the fins because during aggressive interactions these are often the places fish nip and attack each other. We had also suggested looking at the sensitivity of the flank of the fish because this is the part of the body usually handled in farm situations. The reviewers who offered their opinions about the scientific merit of our application, however, stressed that it would be more interesting to find out if a sharp object passing through the mouth of a fish would be painful. Clearly recreational fishing was what these scientists wanted to know about, not fish farming. A resubmitted proposal that included investigations of possible pain detection around the face and mouth of the trout was subsequently approved and funded.

When the funds arrived, we interviewed a number of candidates for the research position that came with the grant, and this was how we came to appoint Lynne Sneddon—now at the University of Chester. Lynne had previously worked on crab behaviour and physiology, and during the interviews her skills and dedication stood out. With Lynne on board, our team was complete, and we were ready to begin the search for the nociceptors and their associated nerve fibres.

Given that we were now focusing on the mouth of the fish, we decided to work with the main nerve that serves the areas around the mouth, jaw, and eyes. In fish, and other vertebrates including us, this is called the trigeminal nerve.

In mammals, the trigeminal passes information to the brainstem, not the spinal cord, but the brainstem operates in a similar way—the nociceptive reflex is triggered in the brainstem before the signal is relayed to different brain areas. The trigeminal nerve consists of three different branches that spread out across the face. Each branch delivers and sends signals to and from these sensitive areas. In fact two of these branches may be familiar to you; the mandibular and the maxilliary are the nerves that the dentist numbs with local anaesthetic before he does any major work on the teeth in your upper or lower jaw. The third branch, the ophthalmic, curves up around the eye area.

In mammals and birds there are bundles of fibres inside the nerves. Two types of fibres transmit nociceptive information. One type, called A-delta fibres, is associated with the first sensation of pain. A-delta fibres are typically between 0.002 and 0.014 mm wide and they have a thin layer or coat of myelin around them. This is a fatty tissue that works like an insulator helping the signal within the fibre transmit more efficiently. A-delta fibres can pass the information from the nociceptors quite rapidly, somewhere between 5 and 30 metres per second, which is why they are associated with what we sometimes call 'first pain'. The second category of fibre lacks the insulatory cover, which makes them both thinner and less efficient at conveying their electrical signals. These smaller fibres, known as C fibres, have a width between 0.0002 and

0.003 mm and they conduct their signal between 0.3 and 1.2 metres per second. In mammals, these poorer conducting fibres are connected with 'second pain'. We readily distinguish the two types of pain in ourselves. There is an initial 'ouch' as the A-delta fibres quickly transmit the detection of the damage from a burn (first pain), followed by a slower, duller ache and throb generated by the slower activity of the C fibres (second pain). Like the nociceptors, no one knew whether fish had A-delta or C fibres.

Our work began then by looking for these two types of fibre in trout. After fully anaesthetizing the fish so that they could no longer recover, the trout were treated with a chemical that helps to preserve and protect the nerve tissue to stabilize it. Next the skin and bone around the head was removed to expose the three branches of the trigeminal nerve. Small pieces from each of the branches were carefully cut away and removed. These short sections were then bathed and exposed to a sequence of different types of chemical to further harden the soft nerve tissue and protect the delicate fibres within. The last step was to seal the small sections of nerve into a resin that sets into a firm block, allowing the nerve to be handled without damaging its structure.

Once the nerve sections were fixed and embedded in resin blocks we used a machine that looks a little like a bacon or salami slicer to cut phenomenally fine slices of the nerve tissue surrounded by resin. The delicate slices—just one thousandth of a millimetre thick—were then carefully placed onto microscope slides. We stained these sections with a coloured dye to make it easier to see the different types of fibre within the nerve. Finally we could peer inside the nerves.

As one adjusts one's eyes to look down the microscope lenses one first sees clusters of blue circles—blue because of the coloured dye used in the final stage of preparation. At first it looks rather like crazy paving because there are rings of various shapes and sizes, some quite round, others a little misshapen. These irregular-shaped blue clusters and bundles are in fact the different fibres that make up a nerve and that send the tiny electrical impulses through the nervous system. The blue dye stains the fatty myelin tissue that is mostly around the edges of the fibres; this is why the fibres appear to be blue rings. As we stared at the blue crazy paving we could see, quite unmistakably, both A-delta and C fibres scattered around inside the nerve—so they were there after all.

Comparing the sizes of both types of fibre we found that they were similar to those found in birds and mammals, but one thing that struck us immediately was that there were many fewer of the smaller C fibres than in other vertebrates. Normally one can expect nearly fifty to sixty per cent of the fibres to be C fibres, but in the trout we found they represented only four percent of all the fibre types. This difference was consistent in all three branches of the trigeminal nerve. The significance of the smaller number of C fibres remains a mystery.

At this point, while excited by the discovery, we still needed to show a degree of caution—the presence of these two types of nerve fibre did not prove that they actually transmitted signals connected to tissue damage. To be certain, we had to find nociceptors on the surface of the skin and then, while stimulating them, record electrical activity inside the trigeminal nerve. If we could do that, then we could confirm that the fibres we had seen under the microscope were indeed relaying nociceptive signals.

Trying to isolate areas on the face of the fish where we could detect the presence of nociceptors called for some delicate techniques. First, we had to deeply anaesthetize the trout, so deeply that the fish were completely inert and unaware of what was happening to them. Once the fish were no longer alert, they were carefully placed onto a specially built cushioned cradle that kept them in an upright position allowing us to work on them from above. A water solution that contained more anaesthetic was continuously washed over the gills using a special tube and pump so that the fish could not come round or recover as we took the various measurements we needed. Although the fish were 'knocked out' and unaware of what was going on around them, their nervous system was still functioning, and with the appropriate form of stimulation, electrical impulses could pass along the nerves.

The skin and bone of the brain case were carefully removed from the head of the trout to give us access to the brain. The cerebellum and the olfactory and optic lobes were removed to expose the part of the trigeminal nerve where all three branches come together, the trigeminal ganglion. To keep the ganglion moist during the experiment, a small amount of paraffin was poured over it before a recording electrode was carefully pushed into it. To locate the position of skin receptors on the face of the fish, a fine glass probe was gently applied in different places. When the probe touched a receptor an electrical signal was detected in the ganglion. The skin of the face has many different kinds of receptor, only some of which will be nociceptors, so as different receptors were isolated they had to be tested to see whether they responded to noxious stimuli.

After finding the A-delta and C fibres within the different branches of the trigeminal nerve we expected to find evidence of nociceptors, so it came as no surprise when we confirmed their presence. We were able to isolate and test fifty-eight different receptors scattered all over the fish's face and snout. We used three kinds of noxious stimulation to test the receptors: touch, heat, and chemical. Sensitivity to touch was measured using von Frey filaments. These are fine hair-like strands of metal that can be applied with a carefully controlled amount of force to a specific contact point on the skin. Sensitivity to temperature was measured by shining a narrowly focused quartz light that could be heated to specific temperatures. Finally, to investigate the receptor's sensitivity to a noxious chemical we applied a tiny drop of weak acetic acid, more commonly known as vinegar. To check that the receptor wasn't accidentally triggered by the mechanical action of dropping the vinegar solution onto the skin, we also applied a similar sized drop of water as a control. The water never generated a response.

Of the fifty-eight face receptors tested, twenty-two turned out to be nociceptors. When the noxious stimuli were applied, we were able to record rapid bursts of firing in the trigeminal nerve. Some of the receptors responded to all three stimuli—heat, touch, and chemical—but others were more specialized and responded, for example, to just two of the stimuli, such as touch and temperature.

Once stimulated, the nociceptors on the face trigger an electrical signal that passes through the trigeminal nerve to the brainstem, and then onto the brain. We didn't know at that point where in the brain the information was processed. There were a number of potential places—many of which process similar types of information in birds and mammals. Tracking a signal within the brain isn't easy, but something that can more readily be measured is the end result—changes in the animal's behaviour. So the next phase of the work involved finding out whether an injury that triggers the nociceptors in the fish's skin also goes on to affect how they behave.

We decided to use noxious chemicals to trigger the nociceptors. The advantage to using a chemical like an acid solution is that it can be carefully placed in known quantities and concentrations on the same spot on different fish.

This precision would allow us to be confident that we were reliably repeating the same 'injury' to different fish. If several fish showed the same changes in behaviour, we could be certain that the acid was responsible. Or could we? In fact, this evidence is not quite enough on its own. Fish do not cope well with being handled, so we would need to show that changes in behaviour were not actually caused by the handling process itself. To rule this out we would need to include two further groups of fish in our experiment that would be handled just like the test fish; however, these so-called 'control' fish would not be treated with the chemical. Together the test and control fish would allow us to see what changes in behaviour arise because of being handled, and whether any additional changes arise because of stimulation with a noxious chemical.

In the end we selected two different types of chemical stimuli, bee venom and vinegar. We decided to use both of these because they generate different types of pain response in other animals. Bee venom is a toxin that creates an inflammatory response, causing localized swelling around the affected area. When you are stung by a bee, the first thing you experience is the sting itself, a sharp needle-like sensation as it pierces the skin, but shortly afterwards the skin around the site of the sting begins to feel most uncomfortable. A cold compress is one of the most effective ways of relieving this irritation. The localized pain that the bee venom generates is largely associated with the swelling it induces. Vinegar also produces a local irritation, but it doesn't generate the same inflammatory response as a bee sting. Accidentally spilling vinegar into an open cut generates a sharp nip and sting in the skin. This is caused by the acidic ions in the vinegar triggering the nociceptors around the cut.

Deciding which chemicals to use was relatively straightforward, but it was harder to decide what behaviour to observe. When humans experience pain, we often start to breathe more rapidly and our hearts beat faster so these were two possible measures we could start to monitor. For the purposes of our experiment, we wanted the fish to swim around unhindered, which made it awkward for us to measure their heart rate. But breathing rate can be easily measured by counting the number of times the gill covers open and close. We measured gill beat rate in fish at rest and then compared it to that after fish had been treated with vinegar or bee venom. By monitoring by how much and for how long the beat rate is elevated, we could put values on how strongly a fish reacted and for how long it was affected.

Another common response to stress and pain is a drop in hunger level. Stressful experiences often suppress our desire to feed. To see whether this was also true for fish, we devised a simple training programme to condition the fish. When a light was switched on above their tank, the fish were given a few food pellets dropped through the centre of a plastic ring attached to one of the glass walls.

After a few days the fish automatically swam to the food ring when the light was turned on. This learned response gave us a measure of how motivated the fish are to feed—a fish that isn't hungry shows no interest in the light or the food ring.

We ran these trials with individual trout, one fish to a tank, so that we could observe their responses to the treatments without worrying that other fish might be influencing behaviour. Trout are typically rather wary, and in the lab they usually head for cover the moment they see a person. To make sure that the test fish were as calm and as undistracted as possible we covered the walls and lid on one side of each tank to create a dark half and a light half. The food ring was placed in the bright section so that to feed, the fish had to move into the lighter half of the tank to reach the food pellets. We also put up a dark curtain between us and the fish tanks—this was to further minimize any effects that our presence might have on the way the fish behaved. We watched the fish through small slits in the curtain. This way we were confident we could move and make notes about the various behaviours without the fish being aware of us.

Each trial began with measurements of the normal breathing pattern of a resting fish. We simply counted how many gill beats the fish made over a minute while the fish was unalarmed and resting. Shortly after this the fish were gently netted and then lifted out of their tank and placed into a tank of water that also contained an anaesthetic solution. Once in this solution the fish became calm and quickly stopped swimming, their breathing rate slowed a little and eventually, as they began to feel the effects of the anaesthesia, they lost their ability to balance and slipped onto their side. To check the fish was now unresponsive, we gently and carefully applied a needle to its tail. If a fish was still alert it flicked its tail and maybe even briefly righted itself, but if there was no response to the touch, the fish was ready to be handled and we considered it to be fully anaesthetized.

At this point, each fish was allocated to one of four possible groups. One group received the bee venom, and a second the vinegar. Fish in these two test groups were moved from the anaesthetic solution onto a moist wet cushion, where they were given a small injection of either bee venom or dilute vinegar just under the skin around the mouth. The fish were then quickly returned to their home tank. The fish in the two control groups were netted, anaesthetized, handled, and then either returned back into their own tanks or given a small injection of saline (saltwater) before being returned to their tanks.

Straight after receiving their various treatments, the fish were given a short time to recover before we started to make our observations. We then monitored each fish every 15 minutes. Gill beat rate was measured and the fish's motivation to feed was monitored by switching on the light and dropping a bit of food into the tank.

All the trout clearly found the handling and the anaesthesia stressful—when they came round, none of them showed the slightest sign of hunger despite having no food for a day. And just by looking at the speed with which the gill covers beat back and forth we could see that their breathing rate was greatly increased. Some fish, however, were clearly more stressed than others. The fish treated with either bee venom or vinegar had a much more exaggerated response. Gill beating increased from about 50 beats a minute in the resting fish to about 70 beats a minute in the fish that were just handled, or those given a dose of saline, but the beat rate of those fish given bee venom or vinegar had increased closer to 90 beats a minute—their breathing rate had almost doubled.

As the fish came round, they mostly sat on the bottom of the tank, resting on their two front pectoral fins and on their tail. The trout that had been injected with bee venom or vinegar sometimes rocked from one side to the other, gently rolling between their two pectoral fins. Occasionally these fish also made darting movements. Several of the fish treated with vinegar rubbed their snouts on the glass walls or on the gravel at the bottom of the tank. It seemed that the stinging action of the acidic vinegar was irritating in the fish's snout. Rubbing it against the tank walls or the gravel might be their way of trying to relieve the irritation. We humans often respond to the nip and sting of vinegar or lemon juice in an open cut by pressing and rubbing the affected area.

We kept watching each fish every quarter of an hour. Nothing much changed over the first hour but then, little by little, we began to see some subtle changes. Fish that had just been handled and those treated with saline were the first to slow their gill beat rate back to about 50 beats a minute. At about the same time, those fish also began to respond to the light switching on. To begin with, they didn't always approach the food ring but they seemed more alert when the light came on. About 80 minutes after they had been handled or injected, fish from both control groups were swimming up towards the food ring and feeding on the drifting pellets as they sank through the water.

But the trout that had been given bee venom or vinegar continued to show no interest in the food and their gill beat rate stayed above 70 beats a minute even after the second hour passed. Eventually their breathing rate did begin to decline but it didn't return to the resting level of about 50 beats a minute until almost three and a half hours after they had been initially exposed to bee venom or vinegar. And around that time the fish's motivation to feed began to return. If you have ever been stung by a bee you will have felt the pain for several hours until the local swelling subsided. Together, these behavioural observations seemed compelling; treating fish with bee venom and acetic acid clearly affected both breathing patterns and an interest in food compared to fish in the control groups.

At this point then, we had evidence of the physical presence of the pain detectors (the nociceptors), evidence that these actively detected tissue damage, and that this information was transmitted to the trigeminal nerve when they were stimulated, and that the behaviour of the fish was altered. With the experiments completed, we wrote up our observations in a scientific paper and, following peer review, the paper was published in the *Proceedings of the Royal Society of London* in May 2003. Just as our paper was due to be published a number of press releases were circulated. We were amazed at the response—almost immediately the phones rang and emails flooded in with requests for interviews. Several newspapers wanted to feature the work and fairly quickly we became proficient at giving the journalists their sound bites. The day the paper was published, various radio and television networks wanted to cover the story. The university car park saw one truck with recording crew appear just as another was leaving. By late afternoon, things finally began to return to normal and I left my office to take my children to their weekly swimming lesson. I had just found a seat in the viewing area by the pool when my mobile phone rang and I was told that a Sky News crew complete with a satellite dish were outside my office expecting me to appear live on their 6.00pm broadcast. Having dried the children off, I headed back to the university to be interviewed once more.

The question that we were asked over and over was 'So, tell us, is angling cruel—are the fish in pain?' But we couldn't answer this. Our study had focused on whether trout had specialized pain receptors and how they responded when these were stimulated with bee venom and vinegar. What we hadn't done yet was to show that the fish were perceiving or suffering from pain. The best we could do was to reply, 'Well possibly....' But this failed to satisfy many of those reporting the work, and so in the newspapers and in introductions to interviews we found that words were put into our mouths and we heard how our research had shown that angling caused pain and suffering.

But we knew we had not shown that. A key experimental test still had to be done. To be convinced an animal experiences pain, we had to show that a complex behaviour is affected. It is not enough to demonstrate a reflex response, or a change in a physiological state like breathing rate, or indeed hunger. One could always argue that a fish under stress secretes stress hormones, such as cortisol, and hormones like cortisol are known to suppress appetite. Losing the motivation to feed may just be a result of the fish being stressed and we could be seeing a loss of appetite without the fish being aware of the sense of pain at all. A fish could lose its interest in food without in any sense being said to suffer. Many animals, including single-celled ones, get hungry. Hunger can be a subconscious process.

So we needed to find a complex behaviour, something that requires a higher order cognitive process that could be reliably measured to see how it was affected by a noxious treatment. Trout are very sensitive to new things. If you place a new object into a tank most show strong avoidance behaviour—at least initially, until they recognize that the new object is not a threat. To detect that the object is novel and should be avoided requires that they pay attention to it. Attention is regarded to be a higher order cognitive process; the animal needs to focus on a single thing while ignoring other aspects of the environment. It needs to perceive that something is new. It requires some sense of awareness. Sometimes different pieces of information may need to be integrated as the animal assesses the object or event that is the focus of its attention. How the animal then decides to behave can be altered by its perception of the situation. Attention to novelty then is a relatively complex form of behaviour requiring higher order cognitive abilities. Hunger, one of the measures in the previous experiment, does not need an animal to be aware of that state. But recognizing and focusing on a novel object demands that the animal be cognitively aware. So we designed a further experiment to investigate how avoidance of a novel object was affected when fish were treated with vinegar.

To assess how warily fish responded to an object they had never previously encountered, we measured how close they got to it. If the fish never swam close to it then we considered they were aware of it and avoided it. Conversely, if the fish did approach the object it would tell us that they were less wary of any possible threat that the object might pose. We used this novel object test for comparing the responses of trout given a control treatment of a saline injection in the snout to those of trout given an injection of weak vinegar solution.

Again, the fish were anaesthetized before being given the small injection of saline or vinegar just under the skin. The trout were then allowed to recover from the anaesthesia before a novel object was placed into the tank at a certain distance from the fish's head. We used a brightly coloured Lego brick tower to act as the novel object.

The results from this new experiment were both striking and clear-cut. Fish given the saline showed a strong avoidance response, hardly ever going close to the Lego tower. Trout given injections of vinegar, however, behaved quite differently. For almost a third of the trial time we found that the fish treated with vinegar moved quite close to the Lego tower. They seemed much less fearful of its presence. These fish were not showing the usual avoidance responses. To us these results showed that the vinegar injection was impairing the fishes' attention, as expected if the fish experienced discomfort and pain associated with the vinegar treatment. To be convinced of this we needed to test whether our 'distraction' hypothesis was true. If our interpretation was correct, we ought to be able to increase the fishes' attention and wariness to the Lego tower by giving them some form of what humans would call 'pain relief'.

If their apparent distraction was pain-based, then by relieving the pain symptoms we should expect the avoidance response to return.

So we repeated the whole experiment as before, but this time, in addition to either a saline or a vinegar injection, all fish received a small dose of the opiate morphine. We chose morphine because it is a broad spectrum painkiller. Fish have an opioid system and so we had every reason to think these drugs used for pain relief in humans could also work on fish. And indeed, just as we had predicted, vinegar-treated fish given morphine showed a more normal avoidance response. In fact we no longer saw a difference in the avoidance behaviour between the trout treated with saline and those given vinegar. By providing pain relief we saw the same level of wariness and avoidance of the novel object in both groups of fish. Now we really were excited. This result was the most direct evidence yet that fish really perceive and experience pain. Giving the fish an injection of a noxious substance distracted its attention, but when pain relief was given, the ability to focus its attention increased again. For this to happen the fish must be cognitively aware and experiencing the negative experiences associated with pain. Being cognitively aware of tissue damage is what we mean when we talk about *feeling* pain. This is a crucial piece of evidence for fish, but showing that fish have feelings and emotions is difficult, as I will explain further in chapter 4.

As we published our results in scientific journals, we became aware of the research done by biologist Professor Chervova and colleagues at Moscow State University. Unbeknown to us, some of their earlier work had actually been published before ours, but because it was in Russian, in Russian journals, we had missed it. The timing was uncanny: at the same time Mike and I were writing and discussing our research proposal, the Russian group had just started to collect data. On learning about their work we were both relieved and pleased to find that they reported findings very similar to our own. Chervova and colleagues had investigated the relative sensitivity of different parts of the fish and found that nociceptors were widely distributed across the bodies of trout, cod, and carp.

The Russian researchers found that the area around the eyes, the nostrils, the fleshy part of the tail, and the pectoral and dorsal fins were most sensitive to stimulation. And while there were nociceptors elsewhere on the head and the body, these were generally less sensitive. Using the common carp they had also examined how the opioid system modulated nociceptive responses. A special chamber through which water could flow was used to confine carp so their movements were restricted. Electrodes were then introduced into the tail fin tissue, into an area some of their earlier work had shown contained nociceptors. The responses carp made to short, mild electric shocks were recorded an hour before and then every 5 minutes for 90 minutes after the fish were treated with a drug called tramadol, a painkiller that targets a specific type of opioid receptor. Chervova and colleagues found that tramadol decreased the sensitivity of the fish to the electric shocks within 5 to 15 minutes of receiving the drug. They tested different doses and found that the higher the dose, the faster the pain relief took effect. This experiment confirms that carp have an opioid system that works similarly to the one found in mammals.

Since then a number of other researchers have also investigated fish pain perception. In 2005, Peter Laming and a PhD student Rebecca Dunlop based at Queen's University, Belfast in Northern Ireland, showed that a pinprick on the flank of goldfish or a trout generates a nociceptive response that can be detected in the spinal cord. Specifically, they found a sensitive region of tissue just behind the gill cover. This area was further investigated to find out whether the pinprick generated nerve activity not just in the spinal cord, but also within the brain. The fish were tested while confined within a small Perspex chamber to minimize movement. Under anaesthesia, a small section of skull was removed to allow a recording electrode to be placed into different areas of the brain. A sharp pin or the soft bristles of a paintbrush were then applied to the skin just behind the gill cover. The recording electrode monitored electrical activity from three different areas within the brain. These intricate experiments showed that both goldfish and trout could detect the pinprick and that the signal it generated was relayed to different areas within the brain including the telencephalon, or forebrain. When the effects of the pin and the brush were compared, it was found that there was a much stronger response to the pin in goldfish, but curiously trout responded in a similar fashion to both the brush and the pin. Why the trout are less discriminatory is not known. Overall, these results are very important because they show that the forebrain of the fish is involved in the response to a pin prick—the forebrain is the place in birds and mammals where higher order information processing occurs.

Laming has subsequently performed a number of experiments investigating changes in the behaviour of goldfish and trout to electric shocks. Using electrodes implanted into the sensitive area behind the gill cover the researchers were able to deliver mild electric shocks to the fish. The electrodes were attached to the fish with long, lightweight leads so that the fish could still swim around their tank. Each time the fish moved into a specific area in the tank the researchers would deliver a shock. Very quickly, the fish learned a map of the tank and clearly avoided places where they experienced a shock. These results compliment the work we did with the novel object by showing that higher order behaviours, such as spatial learning and memory, can be affected by noxious stimuli. Here the fish must stay alert to keep out of the zone that gives them the shocks.

A research group at the Vet School in Oslo, Norway, working with salmon have also investigated nociception and pain perception. PhD student Janicke Nordgren, her supervisor Tor Einar Horsberg, and collaborators Birgit Ranheim and Andrew Chen have found that salmon given mild electric shocks at the base of their tail convey this information to their forebrain. This Norwegian study was quite similar to the experiment performed by Peter Laming and Rebecca Dunlop—a single recording electrode placed in the forebrain of an anaesthetized salmon recorded activity in response to an electric shock. Using four different intensities of electric shock, the Norwegian researchers discovered that the nervous system of the salmon makes distinctions between the strength of the noxious stimuli. As the intensity of the shock increased so did the time-course of the response recorded in the brain.

After finishing the project with Mike and myself in Edinburgh, Lynne Sneddon left for a new position at the University of Liverpool. Still working on aspects of pain perception in fish she has made some significant new observations. With Paul Ashley and Catherine McCrohan, Lynne found that the eyes of trout contain nociceptors. Using techniques similar to those we had used to locate nociceptors on the face of the trout, Lynne and her colleagues found 13 different receptors spread across the cornea. Given the delicate nature of the eye, it is not too surprising to find that trout, like mammals, also protect their cornea with the presence of nociceptors. With PhD student Siobhan Reilly and collaborators Andy Cousins and John Quinn, Lynne used state-of-the-art technology to monitor changes in gene activity in the brain of trout and carp in response to an injection of dilute acetic acid in the snout of these fish. Several genes were found to be active, including some known to be involved in mammalian nociception.

These different studies from around the world show that fish can generate nociceptive responses from several parts of their body, and that specific kinds of stimuli generate more than the nociceptive signal, because information is passed up to higher order brain structures such as the forebrain. The different experiments have also shown that nociception and pain perception occur in several different species of fish including trout, goldfish, the common carp, salmon, and cod. Some experiments examined changes in simple behaviour when fish experienced bee venom or vinegar—breathing rate rapidly increased and appetite was suppressed. Importantly, other experiments demonstrated that higher order cognitive processes including attention and spatial behaviour were significantly altered by noxious stimulation. Impaired cognitive ability caused by noxious stimulation was relieved, however, when fish were treated with morphine—a form of pain relief. So fish respond to noxious stimuli in ways that indicate they perceive pain. The next question that needs to be tackled is whether the fish suffer from that experience.

Suffer the Little Fishes?

Jeremy Bentham, an English philosopher famously wrote, 'The question is not can they reason, nor can they talk, but can they suffer?' What is remarkable about this quotation is that it is over 200 years old—Bentham, social reformer, advocate for utilitarianism, and one of the earliest proponents of animal rights, wrote about the moral status of animals as long ago as 1789.[1] Yet, the quotation poignantly captures the very essence of why modern society is concerned with the welfare of animals. Suffering is an unpleasant sensation; that's why we usually try to

---

[1] Jeremy Bentham, *The Principles of Morals and Legislation* (London: Menthuen, 1789).

avoid it. It can mean different things to different people, but ultimately suffering is a negative form of emotion. It's the possibility that animals may experience something akin to what we humans recognize as suffering that underpins why we take an interest in animal welfare. If animals are incapable of suffering, then why should we care about how we house them or the way we slaughter them?

Feeling pain *is* an emotional experience. There is good evidence that this is exactly how we deal with pain. Images taken from our brains as we experience a painful event reveal a great deal of activity in the areas associated with emotion—the limbic system and, particularly, a structure called the amygdala. To find out whether animals feel pain, we need to design experiments that let us glimpse inside their minds to test what the animals are mentally experiencing. This might sound far-fetched, but in fact it is what experimental psychologists have been doing with animals over the past half-century as they have probed different aspects of animal cognition. It turns out that animals often store information in different forms of representation. These representations provide us with a way of figuring out how the animal is perceiving and considering certain events or experiences. Do fish experience feelings and if so, does this give them the capacity to suffer? We have already seen that fish have a functioning nociceptive system that transmits signals that can quash complex behaviours. Next we need to find out whether fish experience the negative sensation of pain—do they really *feel* the pain?

Many people have argued that suffering is only possible in animals that are conscious because you need a conscious brain to generate sentience—an ability to generate feelings that permit the mental experience of discomfort. If we are to discover whether fish can suffer, we need to ask whether fish are conscious, feeling animals. That is what this chapter is about. It is not an easy question, not least because the interpretation of what we mean by 'consciousness', 'sentience', 'feelings', and 'experience' are open to debate, even in humans. Finding evidence of animal consciousness is a problem that researchers working on animal welfare have previously had to tackle. While never fully resolving the issue of what consciousness is, they have gathered sufficient evidence to support the intuition that most of us hold that the animals we farm, such as cattle, sheep, pigs, and chickens, and many of the animals used in research facilities, such as rats and mice, are sentient and so can experience emotions such as pain and suffering. Recognition that these animals are sentient has, in many countries, resulted in our modifying the ways in which we care for and interact with them. A good starting point to determine whether fish are also sentient might be to borrow the ideas and methodologies previously used to explore sentience in these domesticated and laboratory animals.

The mystery presented by consciousness has attracted many people from a broad range of disciplines: philosophers, psychologists, neuroscientists, cognitive scientists, and more recently researchers in cybernetics and artificial intelligence. This diverse group of minds has tried to explain what consciousness is. In humans we consider that consciousness underlies our thoughts and sensations. It affects our moods and emotions, provides us with an ability to integrate complex information to make informed decisions, and it gives us self-awareness and the ability to communicate through language. Pinning consciousness down isn't easy—especially when we see that it plays such divergent roles. But these multiple roles are helpful, because they provide us with specific processes or categories of consciousness. This approach to studying consciousness is sometimes referred to as a modular view—processes can be split up into different forms and categories.

The different categories or modules of consciousness can be thought of as pieces of a jigsaw puzzle. If a sufficient number of the puzzle pieces are identified, then this could provide us with evidence that animals also possess a form of consciousness. This is a somewhat speculative process—it has to be—but searching for subcategories of consciousness-like components could be a promising route forward. If we think about the processes that will have allowed our own consciousness to have evolved, then it seems likely that other animals will possess simpler forms of these processes. This line of reasoning may extrapolate to other mammals, and perhaps to certain species of bird—but could it also extend to fish?

It is important to recognize at this point that the types of consciousness we may discover in other animals are likely to be very different from our own. The greater degree of sophistication of the human brain compared to say a bird or a rodent, and certainly a fish, is incontrovertible. The human brain is unrivalled in its complexity. Our brains consist of approximately 100 billion neurons, providing enormous potential for processing information. We assume a fish brain will have fewer neurons but we don't actually know how many—and given the diversity of fish brains this number could be very variable. In searching for consciousness in animals we need to appreciate that we are probably looking for mechanisms or processes that will be more simplistic than our own. This is a critical point, because the arguments that have been raised against the capacity for fish to suffer claim that they cannot feel pain the way that we do. This interpretation seems perfectly reasonable. How could fish with their simpler brains possibly feel things in the same way that we do? But remember Nagel's discussion of bats: just because animals experience the world differently does not mean that they will be completely devoid of emotion, or without a capacity for some form of suffering. As we search for conscious-like states in animals, we could be searching for a relatively simple system, but nevertheless something that has a clear impact on the animal's behaviour and well-being. It will influence what decisions are made and may change how the animal perceives the situation it is in.

There have been many attempts to categorize the concepts underlying consciousness, but for the purposes of this chapter, where we need to find elements of behaviour that resemble aspects of consciousness, I find it useful to consider three categories as proposed by Ned Block, a philosopher of psychology and cognitive science at New York University. He has described 'access consciousness' as the ability to think about or describe a mental state either current or associated with a memory—we can introspectively think about information and we are aware of our own thoughts about that information. It is similar to something that had previously been called 'primary consciousness', which is the ability to generate a mental image or representation into which you can combine diverse pieces of information and then use this integrated knowledge to guide your behaviour and the decisions that you make. For instance, you will have a pretty good mental map of the town you live in and this allows you to compute a novel route home from somewhere you only rarely visit. Block's second category is 'phenomenal consciousness'—the experience of sensing what is around you and the feelings and emotions generated by what you detect. This has been called the 'hard problem' of consciousness because it seems impossible to conceive of a mechanism within the brain that could generate phenomenal consciousness—the feeling that lets you know you exist. This idea also captures the essence of what we mean by 'sentience'—an ability to subjectively feel and perceive your environment. Block's third category is 'monitoring and self consciousness'—the experience of thinking about your own actions, the ability to play these out mentally so that we can reflect on a situation and consider different potential scenarios. It is rather like 'extended conscious-ness' that has been considered a 'higher' or advanced capacity that permits self-awareness and, in humans at least, the use of language to communicate with one another. It may seem like a tall order, but let's try to look for these capacities in fish and see whether sufficient pieces of the jigsaw are present for a picture of fish consciousness to emerge.

To start this process, let's explore different ways in which fish create mental representations and how they acquire these. These ideas fit into Block's first category of access or primary consciousness. We need to determine how fish perceive information about the world around them and how this affects their behaviour. Experimental psychologists have been taking just this kind of approach since the 1960s. In fact a number of pioneering researchers even tested goldfish alongside pigeons and rats in psychology laboratories. This trio was used to compare cognition along a scale of relative brain complexity. Associative learning tests, where an animal's actions are either rewarded or punished, allow us to see how animals categorize different objects or events. And the animals in question are remarkably obliging in terms of learning to press levers with their paws, peck at keys with their beak, or nudge paddles with their snout. These actions can be used to gauge and monitor the choices the animals make and their responses to different kinds of stimuli and information. For example the rat, pigeon, and goldfish can be trained to expect a food reward when it makes a correct choice, or be punished if it selects the wrong one—the punishment could be that food is withheld, there might be a long delay before the next trial starts, or the animal may be given a brief, low-level, electric shock.

It is remarkable what these kinds of straightforward experiment have revealed about how the animal mind works. Experimental psychologists have discovered an impressive degree of sophistication that permits animals to make complex decisions. Such work has, for example, revealed staggering visual memory capacities in pigeons—they can learn to recognize hundreds of different picture images, and can accurately remember these over several years. While this feat of memory appears extraordinary and indeed rivals human performance in this test, it leaves us wondering why the pigeon brain should be so adept at visual memory tasks. Field experiments have found an answer. By following pigeons as they fly home from release sites several miles from their loft, researchers have found that pigeons learn and remember visual landmarks to help them find their way home. Some pigeons even learn to use smells as landmarks. The birds combine these cues to generate an internal map of the area around their loft and this helps them reach home even when they are released from places they have never been before. Once the pigeons are in the air they circle around the novel release point until they spot a landmark or a place in the distance that they recognize from their own mental map. They then head towards this landmark until they are back within familiar territory and can locate their home loft. Experiments such as these demonstrate that animals do have and use mental representations, much like we do.

Investigating the role that mental or internal representations play in animal learning and memory could reveal whether animals have a capacity for access or primary consciousness. Even fish need to find their way around the environment they live in—so we can use the choices fish make about where to travel or which route to take as a basis for investigating how they store spatial information. Mental representations are not limited to spatial maps, as studies of social recognition in group-living fish have shown. Recognizing different individuals is important for a structured way of life when a hierarchy influences how individuals interact and behave with each other. Learning and remembering who is more dominant to you will be a valuable skill if you wish to avoid or reduce conflict. Linking the identity of different group members with information on their dominance and fighting ability is cognitively demanding and there are only so many individuals that can be remembered. Typically then, as a group gets larger it becomes difficult to recognize each individual member—at this stage the hierarchy starts to break down. Spatial cognition and social recognition in small groups therefore provide us with two examples to investigate how fish use mental representations—the first is a representation of the environment while the second represents an individual's relationship with others.

One of the reasons spatial behaviour is so well studied is that it is relatively easy to construct mazes and watch how animals learn to negotiate a safe path through them. Once a route has been learned, the shape and features of the maze can easily be altered to put different types of cue into conflict with one another; we might modify the geometry by stretching the maze, or move the position of a coloured wall to alter a landmark. Watching how the animals respond to these changes tells us about the information that they use to navigate. If they shift their movements or searching behaviour because we have moved a landmark, then we know that that particular cue was important to them. For our purposes, the real advantage to spatial learning and memory tasks is that they often rely on the animals forming and using a mental representation. Demonstrating that fish form mental maps would give us the first piece of the jigsaw puzzle—the piece for access consciousness.

How can we test whether fish form mental maps? Rats are famous for their ability to run along the arms of mazes, but what about fish? Can fish too be trained to swim down arms of a maze, through passages and doorways? It turns out that they can, and in fact they are very good at learning where to make turns and remembering routes that they should take. Mazes are an excellent way to study fish spatial ability, and many species of fish have now been tested in them.

A research team led by Cosme Salas and Fernando Rodruíguez at the University of Seville in Spain designed and constructed a glass tank built in the shape of a plus sign to investigate what spatial information goldfish use. The fish were able to swim freely around the four different arms of the tank and could also see out through the glass walls at the objects that were in the room where the maze was kept. In one experiment, individual fish were released at the end of one of the four arms. When they reached the centre of the cross, they could make one of four choices; swim back down the arm they had just been in, or explore one of the three other arms. The researchers wanted to determine how they learned to recognize the different parts of the maze. To study this, they buried a small amount of food in the gravel at the end of one of the new, previously unvisited, arms to see how long the fish took to find it. Once the fish had found and eaten the food the whole process was repeated again and again to encourage the goldfish to learn about the position of the food. The researchers set the maze up in such a way that there were two ways of learning where the food was hidden. The goldfish could reach the correct arm by learning and remembering which direction to turn, for example, swim down to the choice point and then enter the arm on the left. But alternatively, the fish could look through the glass walls and remember the positions of different types of object around the room housing the maze. They could then use these as landmarks to form a map of the room—so they might learn that they should go to the end of the arm that took them towards the sink with the brightly coloured poster above it.

After several days of training, the fish became proficient and consistently fast at finding the food. To determine which of the two methods were used, the maze was modified by rotating it through 180° with the rotation point being the end of the arm that contained the food reward. This meant that the reward was still in exactly the same position with regards the room, but the 180° shift meant that a fish initially trained to turn left would now need to turn right. In this way the two ways of solving the maze were put into conflict with each other. If a fish was using a turning rule—'Turn left at the centre decision point'—it should continue to turn left even in the modified maze. But if the fish had learned the landmarks and features in the room to define the position of the goal then they should swim to the correct place in the room using the position of the sink and the brightly coloured poster to guide them. The fish were given several trials in the rotated maze to see how they responded to the altered setup. As the fish were tested, it became clear that they were quite resourceful. Some fish continued to turn left just as they had during training, but others found their way to the correct position in the room guided by the landmarks—and some fish were found to use both strategies, using turn direction for some of the trials or finding their way to the correct place in the room on others. So the goldfish were able to use both methods—and the fish didn't necessarily just rely on one solution, some actually learned and remembered the turn direction *and* the landmarks.

Experiments with goldfish in a glass-sided maze allow us to watch and test their spatial ability in a controlled, carefully managed setting, but is there any evidence this is the way fish behave in a more natural environment? Lester Aronson, from the American Museum of Natural History in New York, addressed this by running field trials in a 'natural' maze. He used a tiny species of fish called the frillfin goby. These fish live in coastal areas and at low tide they can be found sheltering in rock pools. But these small oases of water are dangerous places because they provide easy pickings for hungry seabirds. The frillfin goby, however, has developed an ingenious technique to evade its predators. The little gobies literally map out their escape route. When you're next poking around in a rock pool you might see these fish suddenly flip from the rock pool you are disturbing into a neighbouring pool. The remarkable thing is the accuracy with which they jump from one pool to the next. They aren't jumping at random. They jump into neighbouring pools. They know where those safe havens are, even though they can't see them.

Aronson was among the first to report this behaviour. He began to devise ways of investigating how the gobies were so accurately finding their way from one rock pool to the next without stranding themselves on dry rock. He discovered that the fish learn the topography of their coastal area at high tide—while freely swimming around the rocky outcrops when the tide is in, the gobies learn the position of depressions that will form the rock pools as the tide recedes. Somehow these tiny fish memorize the positions of the dips in the landscape. At low tide they retreat into one of the pools, but if they are threatened by a predator then the gobies leap and literally flip and jump from one pool to the next. If the fish are still in danger they leap to another pool—if necessary they can make a series of jumps until eventually these remarkable little navigators get themselves down to the open sea where they can escape the reaches of the would-be predator.

A few years after Aronson had first reported these observations he performed further studies to see whether he could find out how long it took the gobies to map an area. By capturing a few fish and housing them in a specially constructed arena that mimicked a patch with rock pools where he was able to artificially generate high and low tides, he showed that gobies need as little as one experience of high tide to learn the new environment. So these little fish—just 10–15 cm long, with a brain about the size of a small pea—have a remarkable ability for fast, accurate spatial learning and memory. They can create and remember a detailed 3-D mental map of the local topography so effectively that it provides them with the ability to plan a safe escape route. Presumably there is strong selection pressure to develop such well-honed navigation skills, because fish not so accurate at mapping out the network of rock pools could easily end up being eaten or die beached on a rock. Aronson's work demonstrates that gobies generate spatial representations of their environment and then use these to plan novel routes—escape routes they have never used before. On the face of it, it is an excellent example of access consciousness.

Is it possible that there may be an alternative, simpler explanation underlying the goby's impressive navigational skill? Animals often use simpler methods for finding their way around—path integration, for instance, in which an animal keeps updating a calculation of how far and in which direction home is. This simply requires keeping track of how far you have travelled and what direction you are moving relative to home. Each time you change direction you update your internal vector and calculate the distance and direction you need to take to reach home. Desert ants that scavenge on heat-exhausted insect prey on baking hot sand dunes use these kinds of calculations so that once they find their prey they can make 'a beeline' back to their nest. Might this kind of process explain what the fish are doing? Well so far, no one has come up with an adequate alternative explanation for the goby's behaviour. The way the fish can take one leap after another as it works its way down the shore to reach the open water requires that it understands the spatial relationship of the connecting pools. The fact it can take novel routes and decide to leap in the appropriate directions after just one experience of high tide strongly endorses the use of some form of map, a map which it has developed itself and which it can use, in a hurry, to solve a problem it has never had to solve before.

Evidence that fish form mental representations can also be found in studies of how fish interact with each other. When two individuals compete over something that they both want, their dispute can escalate into a fight. But fighting can be costly—physical conflict uses up energy and can result in injury. Finding ways of avoiding direct conflict therefore has its advantages. Some animals have evolved ritualized displays, so that the opponents can size each other up before coming to blows. If there is a clear mismatch between the two, the subordinate animal can back down without having to fight. Sometimes however, an escalation to a fight may be unavoidable, and in these cases it is helpful to know the basic facts about your opponent: how aggressive are they, how strong are they, and what's their track record—do they usually win or lose? Equipped with information about your would-be opponent you can be a better judge of whether to quickly back down and act submissively, or whether it is worth asserting your dominance and being prepared to fight.

Fish that protect nests during the breeding season often have to fight to defend their territory. A recent study, by Logan Grosenick, Tricia Clement, and Russell Fernald from Stanford University, made use of this territorial response in male cichlids. These small freshwater fish from the Great Lakes in Africa frequently engage in fights to determine their access to a territory. Previous work has already shown that various species of cichlid and Siamese fighting fish can remember the individual identities of fish they have fought with and that they also learn to recognize winners and losers from fights they have watched. Control tests have shown that the fish are not just learning to look for submissive signals, but rather they are learning the actual identity and fighting capacity of winners or losers.

Grosenick and colleagues wanted to know whether the cichlids could learn to associate relative fighting ability across a series of neighbouring males. To test this they built a central glass tank that housed a single male, known as the bystander because he initially watched fights but didn't take part himself. This male could watch other fish fighting through the glass walls of adjoining tanks. Five different 'fighting' males were housed individually in separate adjoining tanks. Screens could be lowered between the tanks to control the visual contact between all the fish.

Over a week and a half, bystander males would be allowed to watch their neighbours fight and clash with each other. Daily, bystander fish would effectively have a ringside seat as they viewed their sparring neighbours. Being an experiment, the fights were rigged so that every day the bystander saw male A beat male B, B beat male C, C beat D, and D beat E. Looking at this list of interactions you can spot an obvious pattern; A is dominant to B who is dominant to C and so forth. If I were to ask you to guess the outcome of a fight between A and E you probably wouldn't hesitate: fish A will be the victor. Similarly if I asked you who would win between B and D you would most likely pick fish B. You're able to do this because you are following the logic of the sequence; you are doing something called 'transitive inference'. This means that you understand that there is a hierarchy across individual winners and losers and you use this hierarchical relation-ship to recognize specific individuals. This allows you to predict the likely outcome between pairs of fish that have not previously met or fought with each other. We find this quite easy to do, but interestingly children don't. There is a specific stage in development, at about 4 years of age, when a child can begin to spot the pattern and infer the relevant relationship. So what about the fish then—can they get it right? Well in the cichlid's case, yes, they could. Fish put in as bystanders were remarkably good judges of who would be the winners when A and E or B and D were played off against each other. Grosenick and colleagues worked this out as follows.

To determine which fish the bystander predicts will be the winner or loser, they made the bystander 'Piggy in the middle'—male A is positioned on one side of the central bystanders tank, and male E is put in on the other side. When both screens are raised the bystander is suddenly caught in the middle. His response in these circumstances is to fight. So he quickly needs to decide which fish to approach first. It has previously been shown that in this kind of situation the bystander initially moves towards the weaker fish and asserts his dominance. When the researchers presented pairs A and E or B and D, the bystanders would swiftly move towards fish E and D showing that they rated these individuals as weaker. This example demonstrates that these fish can use logical inference to guide their decisions about which rival is a greater threat. To be capable of this, the fish must have memorized the identity of the different individuals and linked this with information on their previous fighting abilities. When the bystanders are presented with a novel pair of fish they can use their memories and associated information to calculate the likely outcome between the new pair. It wasn't that long ago that transitive inference was thought to be exclusive to humans and indeed humans over 4 years old. Yet here, it seems, we have evidence that even fish can make logical deductions.

Before we totally accept this conclusion, can we find any simpler explanations that could explain the results? Being able to predict the outcome between fish pairs A and E could more simply be explained by the bystanders categorizing fish A as 'winner'—the bystander only ever sees fish A in that capacity. Similarly, bystanders might associate the idea of 'loser' with fish E—because again this is the only role they see fish 'E' playing. So simple associative learning could be the way the bystander cichlids solve this problem. But this cannot be the only answer—because the researchers included the crucial comparison between fishes B and D. Fish B and D have both been seen in winner and loser roles (for example, B loses to A, but beats C). And yet, when pair B and D is presented, the bystander males correctly choose B as the winner and D as the weaker fish.

Another possible explanation might be that winner and loser fish convey their status through some kind of status badge. Other animals, particularly birds, visually display fighting ability and aggressive tendency with a visual signal: aggressive great tits, for example, have a wider black breast stripe compared to less aggressive birds. Previous experiments by Rui Oliveira, Peter McGregor, and Claire Latruffe at Nottingham University in England, however, ruled this possibility out. A status badge would immediately indicate the winner or loser status of fish, but this isn't what Oliveira and his colleagues found. A male Siamese fighting fish can only determine winner and loser status between pairs of males when he has directly watched the pair fighting. So, if associative learning and status badges cannot explain the cichlid results, it seems the only explanation left is that they are genuinely solving the problem because they understand the logical ranking of the hierarchy and they use transitive inference to guide their choices.

So far, we have seen that fish develop concepts and representations of the physical features or the social characteristics of the world around them. These examples appear to fit Ned Block's first category of consciousness—access consciousness. The fish integrate different pieces of information to allow them to make informed decisions about which way to swim or who will be the weaker fish. To achieve this, the fish seem to be creating mental representations of the environment or the situation, and these are then used to plan a route or to infer the dynamics of a new relationship. The first piece of the fish consciousness jigsaw is on the table.

Next we need to search for evidence of Block's second category of consciousness, phenomenal consciousness—experiencing the world by feeling it, hearing it, seeing it. This form of consciousness is tightly linked to sentience—an animal's ability to feel and experience emotion. Two excellent scientific articles have discussed the possibility that fish may be sentient. These were written by Rich Moccia, a fish biologist, and Ian Duncan, an animal welfare specialist, and two students Kris Chandroo and Stephanie Yue, all from the University of Guelph, Canada. Their papers coordinate a wealth of examples that span neurobiology through to behaviour and the authors used these examples to argue that fish have a brain sufficiently well developed to permit sentience. They make a compelling case, and I will use some of their examples to argue that the second piece of the fish consciousness jigsaw exists. Before getting to these examples, I want to briefly consider what we mean by sentience.

To accept that fish are sentient, we need to agree that they have the ability to experience emotions. Edmund Rolls, a professor of psychology at Oxford University, describes emotions as states that occur because an animal experiences something positive and rewarding, or negative and punishing. The nervous system processes these experiences and the outcome is that positive, rewarding experiences encourage the animal to try and achieve the same thing again, whereas in the negative or punishing situations the animal avoids it happening again. We recognize that in our own brains there is a collection of structures known to affect our emotional behaviour—the limbic system. Together, these support processes such as emotion and long-term memory. The dopamine system is closely linked with the limbic system. Dopamine is a neurotransmitter that has several different functions, but significantly it is involved with motivation and rewards, and decreased levels of dopamine have been linked with sensations of pain in humans. It is a very critical part of our own limbic system and we know from previous studies that fish have dopamine receptors, but is there any evidence that fish have something similar to a limbic system?

When you look at the brain of a fish it is quite clear that it is a vertebrate brain; we can distinguish a forebrain with two hemispheres, a midbrain, and a hind brain that connects to the spinal cord through the brainstem. But the fish brain is much simpler in comparison to the brains of other vertebrates. The biggest difference is that in fish the outer layer of the brain, called the pallium, is relatively thin, whereas the neocortex found in mammals has multiple layers to it. When brain activity is monitored in people either consciously thinking about feelings or suffering from something painful, we see that several parts of the neocortex become active. James Rose, a scientist from the University of Wyoming and a strong opponent to the idea that fish feel pain, argues that because fish have no neocortex they are unable to process feelings and emotions. Without a neocortex, Rose concludes that fish are incapable of sentience and therefore unable to experience any form of feeling.

Certainly the lack of a neocortex will prevent fish from experiencing things the way we might, but can we really conclude that fish feel nothing? Rose has written a number of reviews pointing out the major differences between the brain of a fish and those of other vertebrates, specifically highlighting the physical contrasts with the human brain. His reviews have been very useful in that they have broadened the issues surrounding the fish pain debate, and they have encouraged considerable discussion. But there is a significant body of work showing that fish are capable of much more than we have previously believed. For example, the goldfish researchers from Seville have not just worked with spatial behaviour in fish, but over the past decade they have also investigated goldfish brains. What they have discovered is quite amazing. It seems that fish have a specialized area in their forebrain that works very like our own limbic system—for instance, it affects how fish learn about processes that have an emotional basis, such as fear. We have only just discovered this because it took a long time to find out what the forebrain of the fish does. It may have broadly similar appearance to other vertebrate fore-brains, but as researchers began to explore it in more detail it became clear that the organization of the fish forebrain is different. This made it very difficult to predict where a structure such as the limbic system might be.

To try and get a better understanding of how the fish forebrain worked, the Spanish researchers began carefully damaging very specific areas of goldfish brains and then seeing what these fish could no longer do. Bit by bit this revealed that the fish forebrain did contain specialized areas, but when the function of these areas was analysed the picture became confusing. Although the researchers were finding areas similar to those seen in mammals, they appeared to be in quite different places. It was only after carefully observing the way in which fish brains develop from an embryo to an adult, determining how certain areas of tissue specialize and where these end up, that the picture became clearer.

After much painstaking work, the Spanish team reported that compared to their land-dwelling relatives, the fish brain appears to be inside out! Structures that lie buried towards the inside of our brains seem to be on the outside,

towards the front of the fish's brain. This striking difference arises at a key stage in the developing embryo. In a
mammal, the development of the brain begins with a
neural tube. The brain starts out as a flattened plate that
begins to invert in on itself, bringing structures on the
outside edges of the plate in towards each other. In most
fish embryos, the opposite happens, with the edges of the
neural tube destined to develop into brain tissue moving
away from each other and so pulling structures apart and
forward, a process called eversion.

The discovery of this key difference in developmental
processes now allows us to make predictions about where
to look for structures like the limbic system. In our own
brains, and those of other mammals, the limbic system
and various associated structures are on the inside of the
cerebral hemispheres. There are several parts of the brain
involved, but two critical areas in this system are the
amygdala and the hippocampus. These can be seen in
Figure 1. The amygdala is very closely linked with states
such as fear, whereas the hippocampus is associated with
learning and memory, determining the timing and
sequence of events, and in particular spatial learning. If we
watch how the fish brain develops, we see that the process
of eversion pushes the equivalent of our amygdala and
hippocampus areas towards the front and the roof of the
fish's forebrain. This too can be seen in Figure 1.

The Seville team confirmed these different effects of
development by investigating the behaviour of fish after

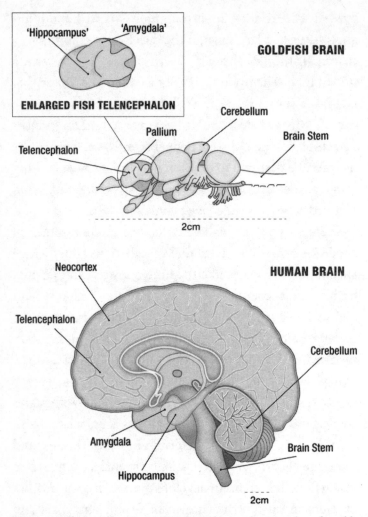

Figure 1 Schematic drawing comparing a human brain to that of a goldfish brain. The human brain is drawn with the external neocortex of the temporal lobe removed to reveal the hippocampus and the amygdala underneath. The Goldfish brain telencephalon has been redrawn to a larger scale on the left to allow the 'hippocampus' and 'amygdala' regions to be labelled.

the hippocampus or amygdala regions of the forebrain had been surgically cut, or lesioned. Once the fish had recovered from surgery their behaviour was analysed. When the hippocampus-equivalent area was no longer functioning, the fish had trouble swimming through a maze that they could readily navigate before the surgery took place. This result is very reminiscent of the spatial learning and memory deficits seen in mammals with damage to their hippocampus. In contrast, lesions to the amygdala area of the fish forebrain made it difficult for them to learn to avoid something unpleasant like an electric shock. The effects of the lesions were extremely specific; fish without a functioning hippocampus could still learn to avoid electric shocks, and fish with the amygdala lesioned could still solve maze tasks. So the lesioning didn't impair learning by itself, but rather a specific form of learning.

So developmentally and functionally there is evidence of a limbic-like area in the fish forebrain. Allied with this area there is also evidence of dopaminergic connections within the fish forebrain. Dopaminergic systems play a crucial role in reward learning and in mammals they are implicated with positive and negative states of mind that form the basis of emotions. Even though the structure and function of the fish equivalent is very much simpler than our own limbic system, the fact that scientists have discovered the presence of similar structures is impressive. It also suggests that from an evolutionary standpoint, the ability to process information with an emotional component may have arisen a very long time ago. The relative simplicity of the fish system—in terms of the numbers and types of connections seen within the brain— might limit the kinds of information they process, but showing that fish have an area specialized to process negative, fear-related stimuli is a major finding.

Again, of course, we must be cautious in how we interpret these observations. Others have previously warned that it can be hard to distinguish between true emotions that we consciously experience as feelings, and the basic physiological processes linked to positive or negative associations. Marian Dawkins, a scientist who has written extensively on the subjects of animal welfare, animal behaviour, and consciousness in animals, proposes that we should distinguish two forms of emotion: the objective form—behavioural and physiological aspects that do not need any conscious awareness, and a subjective form—consciously recognizing and experiencing an event as unpleasant or pleasant. Objective emotion could be described as the body being in an emotional state such as frustration, when the body is awkward and pent up but it achieves this state without consciously thinking about or analysing the frustration. Subjective emotion, in contrast, is feeling what it is like to feel something—so interpreting and being consciously aware of the frustration. Using this distinction we certainly have evidence of an objective form of emotion in fish—but is there sufficient evidence to conclude there is a subjective form? This is less clear.

What role do objective and subjective forms of emotion play? Objective emotion seems to be necessary for animals to learn positive and negative associations—reinforcement learning, where the animal learns to keep responding to things that give a positive reward, but avoid doing things that lead to a punishment. Sea slugs and insects can do this, so objective emotion must be an evolutionarily old phenomenon. When did subjective emotions arise? No one knows the answer to this, but a good guess is that it was associated with the development of the limbic system. If so, the simple form of limbic system in fish may allow them to experience some form of subjective emotion. For the moment, however, this interpretation remains speculative.

But two further examples bear on that speculation. First, recall from Chapter 3 our experiment that showed that fear of a novel object changed when trout were treated with a noxious injection. We proposed that fish treated with acetic acid were less fearful of the tower of Lego bricks—the novel object—because they were distracted by the negative experience associated with the acid treatment. In this context, we are proposing the subjective feeling of the fish—the pain—affects its assessment of the novel object. When we altered the subjective state by providing morphine as a form of pain relief, the fish respond by once again showing the typical avoidance response. This seems to support the idea that fish have a capacity for subjective feelings. But why is this effect any different to the example discussed in Chapter 2 of the snail on the heated plate that slows down its foot-lifting reflexive response when it is given a morphine-like drug? The distinction between these two results is critical. In the snail experiment, the morphine-like drug blocks the nociceptive signal which stops the foot-lifting reflexive response. In the trout example, novel object avoidance isn't a reflexive response because it involves awareness which is a cognitive process—the cognitive process is impaired because of the subjective feeling caused by the acetic acid. Preventing the nociceptive signal using morphine in this case decreases or removes the subjective feeling and this then causes the fish to show active avoidance of the Lego tower.

The second example comes from experiments performed at Queen's University, Belfast, by Peter Laming and his students Rebecca Dunlop and Sarah Millsopp. Trout were trained to recognize that two thirds of a test tank were safe and shock-free, whereas a zone immediately next to this should be avoided because swimming into it induced low-intensity electric shocks. The researchers wanted to determine whether the trout could modify their fear response under different contexts. Trout do not like social isolation, and when fish are kept in individual tanks they show a strong attraction to companion fish even when they are separated by a glass partition. So the experiment was repeated, but this time a companion trout was placed behind a window at one end of the tank, and to get close to it the test fish had to enter the zone that received electric shocks. The first experiment had shown the fish had a strong aversion to this area, but when the companion trout was there, the test fish now moved into and stayed in the zone that exposed it to electric shocks. When the researchers tried the same experiment with goldfish they found that the goldfish preferred to stay away from the companion fish rather than receive electric shocks.

These results reveal a number of things. First of all there are species differences in how motivated fish are to be close to another individual. Clearly this was more important to trout than it was for goldfish. The results also show us that trout change their behaviour under the different conditions. When they are on their own they show a strong avoidance of the zone that gave them electric shocks, but when the choice is to avoid this zone or enter it to be close to another fish then they are prepared to put up with the electric shocks. Trout appear to adjust their choices depending on the context. If we try to explain these observations in terms of objective (behavioural and physiological states) and subjective (conscious feeling) emotions, they seem to more readily fit a subjective explanation. The fish is able to re-evaluate the 'negative' zone depending on the context—companion or no companion—and so makes a subjective choice about where to spend its time.

The presence of a limbic-like area in the fish forebrain, and evidence that fish change the way they view an aversive situation—electric shocks or novel objects—certainly seems to suggest fish have the capacity for subjective emotions. For sure, we have not found a definitive test to categorically show the presence of subjective emotion in fish, so we need to be cautious about how we interpret these results. But we have trouble getting such evidence for any non-human—in fact, we would probably struggle to do this for any human if we could not understand their language. So while we may not be comfortable to attach the phenomenal consciousness puzzle piece to the rest of the jigsaw, I suggest we at least place the piece on the table.

The last puzzle piece and Ned Block's third category of consciousness is monitoring or self-consciousness—the experience of thinking about your own actions, being able to mentally consider different possible scenarios and then modify your decisions on how to act if necessary. To consider whether fish might be capable of this last category I will use an example that involves two different species of fish; an eel and a grouper. An unlikely pair perhaps, but they have developed an impressive alliance. Both live in the waters around coral reefs. The grouper can grow to about 40 cm in length. It has a large mouth furnished with sharp teeth and the ability for fast bursts of swimming that help it to outswim its prey. These attributes make it an impressive predator of reef fish. Moray eels also feed on reef fish, but unlike the grouper they are usually nocturnal hunters and they use their elongated snake-like body to creep around the crevices and holes of the reef to corner their prey rather than trying to chase them down. What do these two species have to do with the search for consciousness in fish? In a recent report, Redouan Bshary from the University if Neuchâtel in Switzerland and his colleagues, Andrea Hohner, Karim Ait-el-Djoudi, and Hans Fricke, described how these two species have developed a way of communicating with one another—and the communication is all about prey.

The grouper and the eel have very different hunting styles, but imagine the formidable team they would make together. When a prey fish that a grouper is chasing seeks refuge in a small hole on the reef, it becomes a sit-and-wait game because the grouper can't follow the fish into the reef crevices. However, it can take some time before the prey re-emerges, and there is no guarantee the prey will reappear in the same place. Coral reefs are a maze of cracks and crevices, so often the prey can emerge on a different part of the reef, far from the hungry grouper. Groupers thwarted this way have learned that rather than sit it out, they can go and get help. They move off in search of an eel, and when they reach their potential hunting partner, who by day rests in a small cave-like crevice, they begin signalling. The grouper does this by vigorously shaking its head making a series of vertical movements several times a second. The eel can choose to ignore the signal, but it often responds by leaving its crevice and then the two swim off together with the grouper leading the way. The grouper takes the eel to the part of the reef where the prey fish was last seen. The eel then starts to explore the reef around that area. Sometimes the grouper even shows the eel where to enter the reef. It does this by almost standing on its head putting its snout into the particular hole where it wants the eel to go and then moves its body into a vertical position. These movements are sometimes accompanied with more head shaking. The now-informed eel can enter the reef and start searching for the prey. On roughly half of the occasions, the eel flushes the prey out from the reef and the waiting grouper quickly darts in and snatches its meal. But this isn't always the case, and there are many times when the eel corners the prey within the reef and gets to eat the spoils itself.

This example illustrates how fish signal their intentions to one another to induce cooperative behaviour. The communication is not a sophisticated language, but the grouper has discovered a way of attracting the eel's attention and the eel has learned what the grouper is signalling. How the two individuals choose to respond to each other varies, but on a significant number of occasions after a grouper has signalled, the eel replies by swimming out of its crevice and then the hunting duo swim off together in search of prey.

The coordinated teamwork between these two different species of fish is striking. There are very few accounts of cooperative hunting in animals, so to find it in fish seems all the more surprising. One of the best known examples of cooperative hunting comes from observations of chimpanzees pursuing red colobus monkeys. Here individual chimpanzees have specific roles that they adopt, such as a driver that chases the monkey in one direction, and blockers who stop the monkey taking certain paths. Using these different strategies a handful of chimpanzees can corner and kill a red colobus. Similarly, there are reports of pairs of hunting hawks that come at the same prey animal from different directions to help to secure the kill. Hawks have also been seen to take on specific roles where one flushes prey, such as a rabbit, out from a shrub where it has been hiding, directly into the talons of the second hawk—the ambusher. The flush and ambush strategy is what the eel and grouper do on a number of their hunts. Hunting dogs, wolves, and lions are known to group hunt cooperatively, and certain sea mammals also choose to hunt together; for instance, dolphins can herd fish into defensive balls that provide a smaller target for them to feed on. All of these examples are cooperative hunting within the same species. The grouper and eel study shows that cooperation is possible even across species.

A joint hunt demands a reasonable level of communication between the individuals in the hunting party, and the ability for the members of the group to understand each other's intentions. A further complication is determining how to share the spoils. In order for cooperative hunting to benefit the different members, they need to be rewarded with access to some of the food. But when there is still food left on the carcass, it is not easy for an animal to learn that it should stop eating to let another animal have a share. The eel and grouper do not have to deal with this complication because their prey is not divisible. On any one hunt only one of the pair feeds, but the dynamics are such that overall both the grouper and the eel have a roughly fifty-fifty chance of getting the prey. This might be aided if the same pair repeatedly hunt together, but the study didn't investigate this aspect.

We will probably never know how the unusual alliance between the grouper and the eel arose. I very much doubt that the grouper saw the eel hunting one evening and consciously decided what a useful partner it would make. A simpler explanation is that at some point, by chance, a grouper may have been hunting in the same vicinity as an eel and on that occasion the eel flushed the prey fish out into the open and the grouper was able to catch it. This might have prompted the grouper to learn that being close to an eel helps flush fish from the reef. Learning the association between the presence of an eel and a prey fish being flushed into the open is not very complicated. The behaviour then developed so that the grouper learned to seek help from the eel, created a signal that the eel could interpret, and so on. Little by little, stage by stage, the behaviour could have become more refined until it resulted in the cooperation we can observe if we go diving in the Red Sea today. It seems likely that the behaviour has subsequently spread among other groupers and eels through cultural transmission. Groupers that watch other groupers signalling and hunting with eels may learn to copy this behaviour so that it spreads through the population and across generations.

However they arose, the grouper/eel interactions are still impressive. What they do and how they work cooperatively requires that both fish learn multiple associations and then blend these together. And these are fish, remember—the animal group that until recently were proposed to be little more than robot-like aquatic animals. Between them, the grouper and Moray eel are behaving in a complex way, but is there anything about their cooperation or alliance that allows us to infer that the fish are conscious or sentient? Monitoring and self consciousness are considered to be the ability to think about your actions and to play out the possible consequences of certain decisions or actions. Does the grouper and eel example fit this category? This one is slightly tricky, because while their interaction seems to illustrate impressive and complex behaviour, their interaction appears to be a one-off. Could we imagine the grouper developing another behaviour that would illustrate they were considering their actions, or thinking through different scenarios? Then again, perhaps this is missing the point. If we consider cooperative hunting as an example of monitoring and self consciousness then surely that's sufficient? Perhaps with that in mind, the Moray eel and grouper represent a perfectly adequate example of a sophisticated, complex behaviour that requires the hunting partners to communicate and recognize each other's intentions, which is all the more impressive for being two species of fish. The final puzzle piece can be added to the table-top picture of consciousness in fish.

So pulling the different threads together, fish really do appear to possess key traits associated with consciousness. Their ability to form and use mental representations indicates fish have some degree of access consciousness. They can consider a current mental state and associate it with a memory. Having an area of the brain specifically associated with processing emotion and evidence that they alter their view of an aversive situation depending on context suggests that fish have some form of phenomenal consciousness: they are sentient. This leaves monitoring and self consciousness, which I argue is in part what the eel and the grouper are doing: considering their actions and pondering the consequences. The grouper is clearly deciding it has no chance to get the prey itself and so swims off to get the eel. The eel is deciding that an easy meal is on offer. On balance then, fish have a capacity for some forms of consciousness, and so I conclude that they therefore have the mental capacity to feel pain. I suspect that what they experience will be different and simpler than the experiences we associate with pain and suffering, but I see no evidence to deny them these abilities, and quite a bit which argues that they will suffer from noxious stimuli.

Over the last three chapters I have built up a picture of what it means to feel pain. We have explored the two parts of the pain process, the initial phase of nociception followed by the emotional part of suffering from the feelings generated by the pain response. We have seen that nociception and nociceptive-like responses are widespread across much of the animal kingdom, including fish. In contrast, the emotional part of pain, the part that generates the negative feeling of suffering is limited to fewer animals. The issues and the evidence are not always black and white, which makes pain in animals a difficult topic with tricky ethical and philosophical implications. However, if we already accept that mammals and birds are sentient creatures that have the capacity to experience positive and negative emotions—pleasure or suffering, we should conclude that there is now sufficient evidence to put fish alongside birds and mammals. Given all of this, I see no logical reason why we should not extend to fish the same welfare considerations that we currently extend to birds and mammals.

Drawing the Line

In 2000, Chilean artist Marco Evaristti was invited to set up an exhibit at the Danish Trapholt Art Museum. His display consisted of ten food blenders containing live goldfish. Visitors to the exhibition were invited to switch the blenders on—the artist explained that his exhibit was designed to make people wrestle with their conscience. The result was that several fish were liquidized and the gallery director was charged, although finally acquitted, of cruelty to animals. Eight years later, the art gallery Tate Modern in London found themselves amidst another fish controversy. Brazilian artist Cildo Meireles created an exhibit that initially contained 55 live, translucent fish, but after 13 weeks on display nearly a quarter of the fish had expired. Various animal rights groups condemned the use of the live fish, claiming that it was inappropriate to exploit these sentient creatures as part of an art display.

The general reaction to these exhibits is interesting, particularly to Evaristti's rather gory concept. His display gave people the option to liquidize live fish—evidently some chose to do this, but it led to multiple complaints. Many visitors believed that killing the fish in this way was unnecessarily cruel. Most of us have gut feelings about what we think is inappropriate. These same kinds of instinctive feelings tell us that birds and mammals can suffer from the experience of pain—which is why we care about their welfare. Something allows us to identify with these warm-blooded creatures. But if we consider fish, we find that people's opinions are much more variable— some people are prepared to switch the blender on, others are not.

For some time now there has been scientific evidence to support our gut feeling about the welfare needs of birds and mammals. There are numerous studies that demonstrate these animals are sentient and cognitively aware of their actions. For example, certain species of bird store food such as seeds, nuts, and even dead insects in different places around their environment, and then rely on their memory to retrieve the food and eat it. Species such as jays store food, but sometimes they do not retrieve it for weeks, or even months. A study by Nicky Clayton and Tony

Dickinson at Cambridge University found that jays learn that specific food items, such as insects, go bad so the birds take care to remember what they hide, where they have put it, and how long ago it was that they hid it. The birds then choose to retrieve perishable items before the longer lasting nuts and seeds.

In a further study Clayton, with colleague Nathan Emery, also at Cambridge University, showed that jays even understand the intentions of other jays. The researchers showed that birds that have learned they can steal food hidden by other jays become very careful in the way they store their own supplies. If they have to hide food when another bird is around they store it, but then later come back when they are on their own and hide it somewhere else. In contrast, jays that have never experienced pilfering do not try to re-store food. This shows that birds that know food may be taken attribute the ability to steal to other jays. The mental concepts that the jays are capable of in both these experiments indicate that these birds are aware of what they are doing.

Using a different approach, Rob Hampton, of Emory University, USA, was able to demonstrate that Rhesus macaque monkeys are consciously aware of their actions. Hampton showed that monkeys knew whether they had an accurate or a poor memory of an event. The monkeys were shown a picture on a touch-sensitive TV screen for a few seconds then the picture disappeared. In the test phase several pictures appeared on the screen and the monkey had to find and touch the image it had seen at the start of the trial. If it was able to do this correctly it was rewarded with a peanut. If the monkey chose the wrong picture no food was given and there was a 15-second wait before another trial began. By increasing the delay between seeing the initial picture and the choice phase with lots of pictures, Hampton could make the task more difficult and increase the chance the monkey would forget the initial image.

To find out whether the monkeys were aware of the reliability of their memory, Hampton added a further phase to the experiment. Now after showing the monkeys the first picture, there was either a long or a short delay before two symbols appeared on the screen. By pressing one symbol a monkey indicated it wanted to be given a test trial with lots of pictures to choose between, but by pressing the other symbol the monkey indicated it was unsure of the initial picture. If the monkey chose this second option it got a less preferred snack but there was no delay before the next trial began. So by choosing to press the different symbols the monkeys were able to report how confident they were that they could remember the initial picture. The Rhesus macaques became very proficient at this task, signalling when they were certain they could remember a picture and wanted to have a memory test. But after long delays, the monkeys often chose to forego the test phase and settled for the less tasty snack. So the macaques are aware of their own ability to remember something.

Using a similar kind of method, Mike Mendl, Liz Paul, and Emma Harding from Bristol University, England, have studied the capacity for emotion in lab rats. Their work discovered that it is possible to determine whether rats are in an optimistic or pessimistic mood. To test this they kept several animals under unpredictable conditions where the rats experienced changes and disturbance—something they find stressful. These animals were then compared to rats kept in standard, more predictable environments.

The researchers devised an ingenious way for measuring the optimistic and pessimistic state of the rats. The rats were trained to listen to two kinds of tone. One was positive: rats learned that by pressing a lever as the tone played they could get a food reward. The other tone was negative: if the rat pressed the lever when it heard this tone it resulted in a delay before the next trial was presented, but if the rat left the lever untouched there was no delay. Once the rats were readily discriminating the two tones correctly, occasional probe trials were added in between the regular trials. The probe trials had tones that were quite like, but subtly different from the tones the rats had learned to distinguish. The researchers wanted to know how well the rats would generalize to the new sounds: would they perceive them as sufficiently similar to the original tones or would they consider them to be different? They predicted that animals less stressed would be more optimistic and so more willing to generalize, whereas animals that were stressed would be pessimistic and perceive the new tone as different. The rats behaved just as the researchers had predicted: those from the more changeable, stressful housing situation were much more conservative and showed very little generalization to the new tones. More recently Melissa Bateson and her student Stephanie Matheson at the University of Newcastle in England have also reported a similar effect in captive starlings.

Although nobody has yet addressed fish optimism or pessimism, I have argued in the last two chapters that there is now sufficient scientific evidence for us to conclude that fish do have the cognitive capacity to experience emotions and that the grouper and eel interactions indicate fish are also self-aware. Based on these kinds of evidence, I proposed that we should extend welfare concerns to fish. I reached this point by asking whether fish have nociceptors, whether these were activated when a noxious chemical came into contact with the receptors, and whether fish respond in ways that indicate they suffer from this exposure to the chemical. In each case the answer was yes. Asking the same questions of birds and mammals also leads to the same answers. As best we can judge, they too have the capacity to feel pain and suffer—a conclusion that at least in their case accords with our intuition.

But what if we asked the same question of animals such as squid, lobsters, insects, or other animals with no backbone? For many of us our intuition tells us that some animals will not have the capacity to suffer from the feeling of pain, and so it should be possible to draw a line somewhere at one of these animal groups. There isn't sound scientific evidence that such a line exists, and certain religions actively try to reject the idea that animals are different. Jains, for instance, believe that humans, animals, and the natural world are part of an interconnected process of birth, death, and rebirth, so there is no dividing line. Yet most people believe that there should be a cut-off point somewhere. Would we really want to accord welfare consideration to earthworms?

Trying to resolve the relationships between different groups of animals posed interesting challenges for early philosophers and scientists. Even Aristotle in Ancient Greece wanted to find a way of ordering the world into a logical sequence. His views promoted a hierarchical picture in which animals were placed into a ladder of life—the *scala naturae*. Aristotle believed that creatures could be lined up on a scale of perfection, with ourselves, of course, taking prime position. Below us were warm-blooded mammals that give birth to live young, followed by warm-blooded birds that lay eggs, and then cold-blooded animals such as fish. After this came the invertebrates, but these were not lumped together. There were subdivisions between those with shells like the snails and those without, such as octopus and squid. Below animals were plants and below these inanimate objects. This ordered, hierarchical view appeared to make sense of life and the world around us, and it came to influence people's views about the animal kingdom for centuries. During the medieval period,

Western Christian culture adapted and modified the ideas of *scala naturae* into the concept of a 'Great Chain of Being'. The goal was to illustrate the impression of perfection. God was at the top of the chain with all life, starting with angels, forming a hierarchy below. Again the divisions and subdivisions provided a sense in which animals could be ranked, and those closer to the top were credited with a higher value.

The innate feeling that it should be possible to draw a line distinguishing animals that will suffer from pain from those that can't may well have roots in concepts such as the Great Chain of Being. Scientifically, however, we don't have sufficient evidence to know whether we can make this distinction. Animal welfare is built around the idea that animals have the potential to suffer and humanity should work to prevent this where possible. Suffering, as we saw in Chapter 4, is an emotional feeling that involves awareness and sentience. Logically then, we should only care about sentient creatures that have the capacity to experience feelings such as suffering. I have argued that fish are sentient. So can we conclude that the line should fall under fish? Let's first ask whether there is any evidence that invertebrates are sentient.

Certain species of invertebrate are already given the benefit of the doubt in some circumstances. The Canadian Council for Animal Care provides legal protection to squid, octopus, and cuttlefish if they are used in research. In the UK the common octopus is protected under the 1986 Animals Scientific Procedures Act. Squid, octopus, and cuttlefish are all members of the cephalopod group. These animals are recognized for their superior sensory systems and their well-developed brains. Octopus and squid are also considered to be among the most intelligent invertebrates. There are plenty of examples of their intelligence to draw on, but one of my favourites is that octopuses can learn how to open a child-proof medicine bottle to get at food hidden inside. To achieve this, the octopus must learn to push down on the lid at the same time it turns it. The first time it tries to do this it can take nearly an hour to get the lid off, but after repeated attempts it can take the lid off within 5 minutes. Such learning skills, clearly more advanced than that of young children, are impressive, but do we really know whether cephalopods feel pain: do squid have nociceptors and are octopuses sentient? Is the fact that they can solve some problems better than children evidence of sentience? We don't actually know, but interestingly legislators in some countries have decided that they should be given protection anyway.

Scientists have recently started to more formally address whether other groups of invertebrate experience pain. A research team headed by Robert Elwood, a professor of animal behaviour at Queen's University in Belfast, Northern Ireland, has recently published the results from a series of experiments exploring whether crustacea feel pain. Crustacea are a very large and successful subgroup of invertebrates that include crabs, lobsters, prawns, and barnacles. They are also closely related to groups such as the insects and spiders, their external skeletons with jointed limbs and appendages collectively bringing these groups together as arthropods. Elwood's motivation to investigate pain perception in crustacea arose from a question posed by a celebrity chef that Elwood met when the chef happened to be filming a seafood television show in Belfast. The chef wanted to know whether certain culinary practices might be considered cruel. Crustacea are often cooked while still alive: lobsters are dropped into boiling water, and prawns and shrimp are grilled on barbecues, sometimes needing to be pushed back onto the grill as they try to escape the heat. Is it ethical to prepare the animals in this way? Are these creatures suffering by our actions?

To address these questions Elwood and his team worked with prawns and crabs. They did not specifically look for nociceptors, but they noted several kinds of sensory receptors on parts of the outer body and that the crustacean nervous system is, in terms of its size and complexity, somewhere between that of cephalopods and insects. Elwood's research group ran experiments that explored the effects of exposing crabs and prawns to noxious chemicals or to brief, low-voltage electric shocks.

In prawns, the researchers looked at the effects of noxious chemicals brushed onto one antenna. The antennae are paired sensory appendages on the front of the head that detect different kinds of information such as smells and tastes, and there are many sensory receptors along their length. Initially, either seawater or a solution of local anaesthetic, called benzocaine, was applied onto one of the two antennae and then the initial responses and also general levels of activity of the prawns were measured. Prawns have a tail flick reflex that they make in response to threats or surprises. This reflex causes a rapid bend in their abdomen muscles that effectively propels the animal away from the area where it was threatened. Prawns treated with the local anaesthetic flicked their tails much more frequently than did those brushed with seawater. Clearly something about the anaesthetic coming into contact with an antenna was initially irritating. The general movement patterns after the tail flicking ceased were found to be similar across all the prawns, so the benzocaine had been given in a sufficiently localized way that it was not acting as a general anaesthetic. One further difference seen in the prawns brushed with benzocaine was that they frequently pulled the treated antenna through their small pincers and parts of their mouth in grooming motions.

The prawns were then given a second treatment. Now seawater, caustic soda (sodium hydroxide), or vinegar (acetic acid) was brushed onto the same antenna that had been treated earlier. The caustic soda, which creates an alkaline solution, and the vinegar, which forms an acidic solution, were used as noxious chemicals. None of the prawns initially treated with the local anaesthetic tail flicked when the second treatment was given. In sharp contrast, however, prawns that had received seawater in the first treatment now showed strong tail-flicking responses if they were brushed with caustic soda or vinegar solutions. The local anaesthetic therefore blocked the noxious effects that triggered tail flicking—a reflex response that Elwood and his group have described as a form of nociception. The prawns initially treated with seawater followed by caustic soda or vinegar also showed a much higher rate of antenna grooming and they even rubbed their antenna on the walls of the tank. The researchers proposed that this grooming and rubbing, unlike the tail flick, was evidence of an irritation lasting longer than the reflexive, nociceptive response, which indicated that the prawns were experiencing pain.

In a different study, Elwood and colleagues studied the response of hermit crabs to electric shocks. These small crabs get their name from the fact that they use empty snail shells to give them protection. When you stare into rock pools at low tide you sometimes see a snail moving at an unusually fast pace. That's because the shell is now home to a hermit crab and it is the crab, not the snail, scuttling across the bottom of the pool. As the crabs grow they need to locate and move into bigger shells. Sometimes the shells they find are empty and this allows for a quick swap so the vulnerable crab is only between shells for a brief time. Hermit crabs have devised a number of ways in which they size up new shells—they use their claws to feel their way around the outside of the shell, measure the aperture and even lift the shell to gauge its weight. Sometimes the crabs also probe inside the shell to take internal measurements. Empty shells are not always available and from time to time hermit crabs must fight each other in order to gain access to a superior shell. These fights are fascinating to watch because the combatants perform pre-fight displays, sometimes rapping one shell against the other in a rapid burst that sounds almost like a wood-pecker drilling for insects. Remarkably, these tiny crea-tures can integrate several pieces of information about the size and quality of a shell, and even the shell's current occupant, and then use this to help them decide whether to fight.

To study the effects of electric shocks on crab behaviour the researchers collected crabs and empty shells from two different species of snail—hermit crabs prefer shells from one of the species. The team then drilled two small holes into a series of test shells from both kinds of snail. The holes allowed wires to be passed through to reach, or be very close to, the abdomen of the crabs, permitting low-voltage electric shocks to be given. Hermit crabs that were to be tested had their current shells carefully cracked in a vice and then removed before individual, naked crabs were put into small beakers that contained a new test shell with its attached wires. The wires protruding out of the test shell were long enough that once the crab was inside the shell it could still freely move around the base of the beaker. The researchers wanted to know whether, by delivering a series of electrical pulses to the crab's abdomen, they could manipulate the crab's decision to evacuate even if there is no other shell for it to move into. The motivation to evacuate was compared to crabs in wired test shells but where no shocks were given. In another set of tests the team investigated whether the species of the snail shell the crabs were in, or the species of an available empty shell, had any effect on how quickly a crab evacuated the test shell. In these trials, empty shells were added to the beakers 20 seconds after the last shock was given—or at the equivalent time for control animals that didn't experience any shocks.

The results were surprising. After being given electric shocks about a quarter of the crabs chose to be naked outside their shell even when there was no alternative shell for them to move into. This behaviour is unusual because hermit crabs are extremely susceptible to predators when they have no shell for protection. Choosing to put themselves into such an exposed position suggests that the crabs found the electric shocks to be unpleasant. Hermit crabs in a test shell of the preferred snail species were less likely to evacuate these shells than crabs in the less preferred species. In trials where empty shells were available, crabs given electric shocks typically left their test shell for the empty one—even if it was a shell from the less preferred species. In contrast, fewer control crabs evacuated their test shell. The shocked crabs also spent less time inspecting and assessing a new shell before getting inside it.

Elwood and his colleagues suggest that these results can be explained in terms of the crabs making motivational trade-offs that are affected by the bad memory of the electric shocks. For example, under normal conditions hermit crabs are motivated to protect themselves and so they stay within a shell until there is a possibility to swap it for a better one. The experience of the electric shocks changes this normal behaviour and now the crabs are motivated to evacuate their shell even when there is no alternative for them to move into. The decision to evacuate is also affected by the kind of shell a crab is currently in; they remained in shells of their preferred species for longer. The researchers suggest that the crabs' response to the shocks is more than nociception. They are making decisions based on information they gather about the world around them, and the decision to evacuate cannot be a reflex response because in trials where a new shell is provided the shell is added 20 seconds after the last shock is experienced. So the crabs appear to form a negative memory about the electric shocks and the test shell that persists more than 20 seconds. This negative memory is proposed to be equivalent to the memory of a painful experience.

Do the results from these experiments really demonstrate that crustacea experience pain? Elwood and colleagues argue that both crabs and prawns show responses that are more than reflexive, nociceptive responses indicating that a process beyond nociception is going on. The prawns show prolonged rubbing and grooming of their treated antenna, and crabs develop a memory of a bad experience associated with their shell. But at this stage we don't have quite enough evidence to say whether the prawns or crabs perceive these situations as painful. In the previous chapter, I argued that the most robust evidence for showing an animal is capable of feeling pain is to demonstrate that the animal is sentient—does it have a capacity for positive and negative feelings and emotions? At this point we do not have any evidence showing crustacea to be sentient. For example, unlike cichlids, groupers or eels, there is no evidence that crabs base their decisions on self-awareness.

Elwood and his team argue that changes in motivational state indicate that the crabs experience pain. They may be right, but motivational states can vary without animals necessarily being aware of any change. When a mouse is exposed to the odour of a predator for example, it becomes more wary in the way it responds. Its motivation to find food drops and it becomes less likely to perform normal behaviours such as grooming. These behavioural changes happen because of changes at the physiological level—changes in the amount of stress hormone being secreted—they do not necessarily arise because the mouse is consciously aware of or thinking about the predator. Similarly when we have experienced a frightening event our decisions and motivations are altered; we become more cautious, for instance. Yet these changes in our perceptions do not require conscious awareness. If changes in motivational state can occur without the need for conscious awareness it would seem that we have insufficient evidence at this point to decide whether the hermit crabs are in pain.

Elwood and his colleagues have also argued that crustacea have a sufficiently complex nervous system to perceive pain because they have an impressive cognitive ability. Hermit crabs integrate information to help them assess a potential opponent, and they can remember the identity of an individual they have previously fought with for several days. These are remarkable skills for a crab, but they are a long way from abilities such as transitive inference that we saw in fish, and so these kinds of skill do not give us sufficient evidence to conclude that crabs are sentient. Interestingly, our views about pain perception in both cephalopods and some crustacea seem to be based on the fact that these animals have superior cognitive capacities.

While cognitive ability should not be ignored, it cannot be used as evidence on its own. Demonstrating intelligence is one thing, but how smart an animal is does not tell us about its capacity for sentience. There are other animals with similarly impressive cognitive skills that we would not consider to be sentient. Honey bees, for example, are famous for their spatial learning and memory skills and their ability to communicate the location of food sources to their sisters through a dance they perform back at their hive. Do these cognitive skills show us that bees are sentient? No, these skills, although impressive, do not demonstrate that a bee has any capacity for an emotional response. Somehow we find ourselves back amidst the grey areas of how we define sentience and consciousness. For this reason, several animal welfare researchers have proposed that we should leave cognition and intelligence out of the animal welfare equation—intelligence is a difficult phenomenon to define or compare across different groups and it doesn't help determine whether an animal is sentient.

In the previous chapter we explored cognitive abilities in fish, but these examples were part of a larger explanation seeking evidence for different forms of consciousness. Can we apply the same framework to look for evidence of sentience and consciousness in cephalopods and crustacea? When we do this, it seems they come close to reaching some of the criteria, but they fall short of what we found in fish. Certainly octopuses have impressive spatial abilities and their ability to plan routes suggests that they have some form of mental map that could be considered to be an example of access consciousness. From time to time there are natural history documentaries that play wonderful video clips of octopuses crawling through large three-dimensional mazes and they do this with striking levels of speed and accuracy. Octopuses are also regarded by those working with them as remarkably resourceful escape artists, and there are numerous anecdotal stories of octopuses in public aquaria being caught making illicit night-time excursions to seek out prey in neighbouring tanks.

In fish we found further evidence of access consciousness by exploring social interactions and the ability of fish to make logical inferences about social, hierarchical relationships. Octopuses are usually solitary animals so investigating their social behaviour is not really biologically possible or meaningful. But squid are much more amenable to such studies as they do come together to school. Jennifer Mather from the University of Lethbridge in Canada has spent much of her career investigating cephalopod behaviour. One aspect she has focused on is the elaborate skin pigment changes squid use to communicate with one another. Sometimes their whole bodies seem to shimmer and ripple as patterns zip up and down the body and tentacles. Mather and her colleagues have shown how squid can use specific body patterns to assert dominance over other individuals, but we still don't know whether they are able to learn the identity of individual animals, and we are a long way from knowing whether skills such as transitive inference are possible. It is too early to say that these skills are not possible in cephalopods, but rather we need to find better ways of getting the animals to tell us whether they can.

Certainly the adaptability and impressive problem-solving skills found in cephalopods suggests that they lead relatively complex lives, but we currently have too little information to say whether they are sentient. It would be helpful to know whether cephalopods have specialized areas within their brain that operate like the amygdala or the hippocampus in vertebrate brains. Is there a specific part of the octopus brain that allows these creatures to navigate their way around their environment or solve complex mazes? Can we run experiments to explore optimistic or pessimistic emotions in squid? Experiments designed to address these kinds of question could help us formally search for signs of sentience in these creatures in the future. For instance, could a method similar to Rob Hampton's Rhesus macaque study be devised for octopus or hermit crabs to determine whether these animals are aware of the accuracy of their own memories? Experiments like this would get us closer to finding out whether cephalopods or crustacea have the potential to suffer.

The uncertainty surrounding nociception and pain in cephalopods and crustacea shows that we still have a lot to learn about the mechanisms that generate these processes and the stages that lead to the sensation of pain. Perhaps working with simpler, invertebrate animal models will allow us to pinpoint what generates pain sensations after nociception. It is curious that almost all the research done on animal pain to date has been directed at trying to get a better understanding of pain in ourselves. If we genuinely wish to improve the life quality of the animals we rear, then surely we need to understand the mechanisms that generate pain in non-human animals. To provide appropriate analgesics or targeted pain relief, it would help to have a thorough understanding of the processes that lead to the sensation of animal pain, and as with much in biomedicine, analysing these problems in 'simpler' organisms could be very interesting.

So where does all this uncertainty leave us—where should we draw the line? Can we be confident in drawing it under fish, a neat divide between vertebrates and invertebrates? Or is it right to include cephalopods and possibly crustacea? We may not have evidence of sentience in cephalopods, but the absence of evidence does not necessarily mean the absence of sentience. Given the rich behavioural and cognitive skills this group of animals possesses, it seems likely that something akin to sentience could exist. In contrast, crustacea seem to have more limited capacities for behaving and making decisions, but we cannot ignore the fact that their nervous systems are responding to noxious treatments in a more complex way than simple, reflexive responses. I look forward to work on what this means.

There is an interesting consequence of the uncertainty surrounding prawns, crabs, squid, and octopuses: while we haven't been able to confidently draw a line ruling them in or out, we have in some ways strengthened the position of fish. The material covered in this chapter has helped to emphasize the relative value of the evidence we have to support sentience and pain perception in fish. It is as good as anything we have for birds and mammals. Fish, like birds and mammals, have a capacity for self-awareness. And so far there is no similarly compelling evidence for any of the invertebrates.

I wonder what visitors to the Danish Trapholt Art Museum would have made of food blenders containing octopuses or prawns? And if certain members of the public had chosen to switch the blenders on, would this have attracted the same level of concern as the display of goldfish? I doubt the gallery director would have been charged with cruelty. It is little different to cooking live lobsters by dropping them into boiling water. Many questions remain unanswered, but for my money, fish should be grouped on the same side as other vertebrates. The open question is what other groups might also be accorded welfare concerns in due course?

Why It Took So Long to Ask the Fish Pain Question—and Why It Must Be Asked

Why has the question of whether fish feel pain only just attracted scientific interest? What was it that stopped us from worrying about fish welfare before? There are several possible explanations, but it's hard not to think that a major reason we have only just begun to worry about fish welfare is that, well, fish are just so different from us. We view them as somehow primitive. Unlike our reactions towards birds and mammals, we tend to feel less empathy towards fish.

We may view fish as very different, but in evolutionary terms fish are remarkably successful—with an estimated 30,000 different species they show a phenomenal amount of diversity. Mammals have only about a sixth as many species, and birds a third. Fish have adapted themselves into a diverse array of aquatic environments; from the pristine freshwaters of streams and lakes to the depths of the salty oceans and from warm colourful tropical reefs to the frozen waters of the Arctic and Antarctic. Several attributes contribute to their success, but in particular their ability to learn and adapt their behaviour has enabled fish to make the most of the environments they inhabit. Yet, being part of a subaquatic world that we can only temporarily visit makes it difficult for us to relate to fish.

Fish seem paradoxical—they look alien and yet without them we would not exist. They first arose 450 to 500 million years ago. The earliest forms looked very different to what we recognize as modern fish. Ancestral fish had jawless heads. Most lacked fins altogether and their bodies consisted of tough, protective scales that formed thick, defensive armour. These early forms could reasonably be described as primitive. Around 380 million years ago, the numbers of jawless fish began to decline and jawed forms started to radiate. Sharks then appeared, with skeletons made of cartilage not bone. Other forms went on to become the ray-finned fish, ancestors of modern bony fish (teleosts), such as cod, salmon, eel, carp, and sturgeon. Other groups included the lobe-finned fish. These are notable because they gave rise to four limbed terrestrial animals that gradually became less and less reliant on water and eventually they left their aquatic ties behind to exploit what was on offer on land. The resulting evolutionary changes in the body shape and physiology of lobe-finned fish played a significant role in making us what we are.

By way of a brief aside, there is a curious myth that attempts to link us with our piscine ancestors. It has been suggested that during our own early development in the womb we go through a fish-like phase where we have gill slits. This intriguing notion was first proposed in the latter part of the nineteenth century by the German biologist Ernst Haeckel. He believed that as an embryo grows, it passes through a series of developmental stages (ontogeny) that mimic the adult forms of more primitive species, thus re-enacting the evolution that gave rise to that animal (phylogeny). Our gill slits, he proposed, reflect our earlier fish ancestry. Haeckel's hypothesis that 'ontogeny recapitulates phylogeny' raised intriguing possibilities, but it turned out to be completely wrong. Curiously, however, these ideas have a certain amount of appeal and every now and again the hypothesis has re-surfaced as a topic for discussion. Modern biologists have demonstrated that the 'gill slits' are in fact grooves around the developing throat area that go on to specialize into key structures such as the lower jaw, the tongue, parts of the ear and even glands such as the thymus. It is curious that we find it tempting to think there may be a direct link to our fish ancestry. There is a link—an evolutionary one. But it isn't a link that we experience as a developing embryo. We need to go back and trace the fossil history to find the connection, not look at developmental stages that define us as mammals.

The origins of fish are millions of years old, but the fish that swim in the rivers and seas of today are, for the most part, very different to those that began to develop limbs rather than fins. A few old forms still persist—the coelacanth is an extant lobe-finned fish with a quite remarkable history. Fossils of these fish are common from about 385 million years ago and the recent discovery of a fossil coelacanth jaw bone in Australia now puts their history even further back, to around 408 million years ago. Put another way, these fish existed over 170 million years before the dinosaurs roamed the earth. For a long time coelacanths were only known from the fossil record and they were considered to have gone extinct with so many other creatures including most of the dinosaurs, at the end of the Cretaceous period about 65 million years ago—casualties of the after-effects of a huge asteroid that smashed into the Earth. So imagine the surprise and excitement when a trawler off the coast of South Africa caught a fresh coelacanth specimen in 1938. Since that first, astonishing discovery a number of coelacanths have been recovered. It is believed that these fish, little changed in 400 million years, live for up to a century. They spend most of their time in deep ocean waters down to 700 metres. A submersible fitted with a video camera has managed to film these fish swimming. They have a curious way of moving their fins that makes them look like they are paddling rather than swimming, but despite that they are still frequently referred to as a 'swimming fossil'.

Discoveries like the coelacanth continue to promote the idea that fish are old and primitive. It is true that the coelacanth has evolved very little over millions of years. But the modern ray-finned fish, like their terrestrial cousins, haven't stood still. They too have evolved into an extremely successful taxonomic group. The number of described fish species continues to expand. It is currently estimated that more than 250 new species of fish are found each year. Some groups of fish generate new species at such a high rate that they have become the focus for evolutionary biologists interested in studying how speciation happens. The cichlids of the African Rift Valley lakes Tanganyika, Malawi, and Victoria are famous for their remarkable diversity, much of which arose within the past 50,000 years. With different species evolving into almost every conceivable niche within the lakes, and indeed several inconceivable niches too: some species have become specialized at feeding on the scales of other fish, developing specific teeth to pinch off and grind scales, asymmetric jaws that allow the fish to bite from the side, and even mimicking body colouration, permitting them to get close to their prey species. Quite literally, these fish provide us with opportunities to watch evolution in action. Thus while some fish seem old and primitive, perhaps for good reason, others have a modern, highly dynamic side to them.

Ironically, it is partly the terminology modern biologists choose to use that continues to promote the idea that fish are 'less' important. For instance, when referring to animals that are older in terms of their evolutionary origin, biologists often describe things in terms of 'lower' or 'primitive' status, whereas descriptions such as 'higher' or 'modern' are used in the context of animals that have evolved more recently. Indeed I did this above. I found it hard not to, even though I knew it was potentially biasing the perception of fish. Such words are deceptive because they generate the false impression that evolutionarily older animals are simpler and less well adapted. Evolutionary success needn't be measured by how recent or complex something is; it might be by how well adapted and diverse it is, or indeed how long something has persisted. The majority of the existing, modern fish species are in fact young; i.e. they have only recently appeared as a species. Thus it is wrong to describe them as 'old', 'primitive', and 'lower'. They are, of course, the descendents of ancestors that appeared a long time ago—but so are we.

Perceiving fish to be different is only one reason why it may have taken until now to ask the fish pain question. Other possible explanations lie in both the history of scientific research and animal welfare and changes in the way we harvest fish. The science we use to investigate animal welfare is surprisingly young. David Fraser, a Professor of Animal Welfare at the University of British Columbia in Canada, suggests that while our concerns with the way animals are treated and thought of may date back hundreds of years, the science behind welfare only formally began about 60 years ago. Indeed, it is only really within the past 30 years that animal welfare science has been recognized as a significant research area. We may only now be asking questions about pain and suffering in fish because of the infancy of this new scientific discipline where the initial focus has been the welfare of cows, sheep, pigs, chickens, and laboratory rodents.

What was it that prompted scientists to become interested in the way we treat animals? Before the Second World War, agricultural operations in the Western world consisted of traditional family-run farms. These were small scale and were typically dependent on manual labour to work the land and tend the animals. There was a general view within society that the farmers cared for their livestock because they were closely tied to the farmers' livelihood. At the end of the War, however, a transition began that replaced old-style farming with production systems that were much more intensive. Animals that had previously spent large parts of the year outdoors were now confined to indoor facilities. By keeping livestock in windowless sheds and using artificial lighting and temperature control, growing seasons could be prolonged and it became possible to produce greater quantities of meat, milk, and eggs. The human contact with individual animals, however, was lost.

As farming practices changed, some began to question the new production methods. Farming was now an industrialized process. Gone were the small farms and in their wake came large profit-driven companies. The public became increasingly wary of this shift and needed reassurances that the animals confined within these large-scale systems were reared appropriately. For instance, how were animals that were used to living outdoors managing with this new indoor existence: could they still behave naturally in their confined quarters? Did the prolonged growing season affect them? Why were chickens pecking at each other in ways that sometimes led to cannibalism? Why were pigs fighting, sometimes fatally wounding one another? Why did dairy cows have lesions developing on their legs? Did it matter? To answer these questions we needed a systematic approach. This was the beginning of animal welfare science.

To find out how animals respond and cope in high-density environments we needed to devise experiments that could tell us about the effects of intensive farming. Were the animals we intensively farmed suffering because of the way they were housed or handled? If so, could we find solutions that would relieve this? Such questions could be investigated by applying scientific approaches. The results and conclusions have been used to inform managers, regulators, and inspectors. In many countries, this has promoted the creation of guidelines that instruct farm workers on how best to handle, house, and work with the farm animals in their care.

As terrestrial farm production switched to intensive rearing practices, technology also changed the way we harvested fish at sea. Bigger, more mechanized fleets now landed many more fish—where once we might have brought up tens of fish at a time, they were now hauled up by the thousand from great depths. Modern fishing vessels are so effective in harvesting their quarry that humanity has managed to deplete many of the world's fisheries. We have fished to extinction stocks once abundant and seemingly limitless. Others are now only just clinging on to their existence after various fishing bans have been enforced.

As wild fish stocks dwindled, we began to devise new ways of producing fish protein by growing it for ourselves. In the 1940s and 1950s we had the 'Green Revolution'; now we are in the midst of a 'Blue Revolution', where fish are farmed by the tonne. The aquaculture industry has grown exponentially in the past two decades and it continues to grow. Just as intensification of terrestrial farming drew the welfare spotlight, concerned consumers are beginning to question fish farmers about the welfare consequences of rearing captive populations of fish by the thousand.

The scientific study of welfare in aquaculture is new, but it is rapidly expanding. On farms, underwater cameras and acoustics can be used to observe how fish move inside their cages. Such techniques can also be employed to monitor fish responses after potentially stressful experiences, like handling or moving fish between cages. These different kinds of measurement can be analysed to find out how the fish react to different contexts and to compare how different fish species vary in their responses. Methods such as this allow us to address the concept of fish welfare and how it relates to aquaculture. In the next chapter, for instance, I discuss how changes in the way fish are fed can enhance both welfare and production. As we learn more about fish welfare in aquaculture it increases our overall understanding of how our interactions affect fish in other contexts, and this has a number of wide-ranging implications.

Recognizing that there are changes we can make to improve how we interact with fish has generated a degree of unease, particularly among those who fish as a hobby. This brings us to the next reason I believe it took us so long to ask the fish pain question. It is possible that we haven't really wanted to know the answer. Asking the question could take us to places we might not want to go: if fish feel pain, what does this mean for our current practices? Should it affect hobbies that involve fish, particularly angling? These are difficult issues that are hard to resolve and one way to avoid discussing them is not to ask about them in the first place. But as I described in Chapters 3 and 4, researchers have gone ahead and asked the question anyway. The proverbial genie is out of the bottle.

Of the many people with whom I have discussed the fish pain debate, the anglers—who in many ways are the people who know the fish best—are the most wary. They are concerned that catching fish using a hook may be perceived as cruel. It seems to me that many anglers have wanted the fish pain debate to go away: they don't want to know whether fish feel pain because finding out that they do may require them to justify their sport. Some of their concern is warranted because animal rights groups are turning their attention to fish and are beginning to campaign against angling and other forms of sports fishing in an attempt to bring about a ban. Anglers are aware that evidence demonstrating fish can suffer pain gives animal activists empirical support for their cause. Paradoxically, many of those who choose to fish do so because they are quite passionate about these aquatic animals. Presumably with potential suffering in mind, many anglers opt to use fishing gear that minimizes the time fish are handled out of water, or quickly put fish 'out of their misery', or intentionally remove barbs from hooks allowing these to be removed from the fish both cleanly and quickly.

Campaigns for animal welfare or for animal rights are nothing new. As the English Enlightenment was underway in the 1700s, philosophers were exploring the basis of ethical behaviour. While much of this was directed at human interactions, it also included discussion of the proper treatment of animals. The way many animals were dealt with at that time would be regarded as quite shocking today. Animal baiting was common and took a number of different forms. Bears or bulls were tied with sufficient slack in the chain or rope to allow a certain degree of movement, but not enough for escape. The animals would then be goaded until they were enraged and would lash out or attempt to take a run at their tormentors, actions in which they were constrained by their chains. Sometimes the animals were blinded to further disadvantage them in this gory spectacle. In bull baiting, dogs would be encouraged to bite and catch the bull by its nose. In a different kind of 'sport', chickens were buried so that only their head remained above ground and people then took turns to take a swipe at it with a long stick with the intention of eventually decapitating the bird. Today such treatment of animals is considered unjustifiable, but two centuries ago it required 35 years of debate before a law banning activities such as bull baiting was finally passed in England. Early campaigners for animal rights were slowly changing the way society perceived and treated animals—although there are still some parts of the world where such sports continue today.

In the first half of the nineteenth century there were debates about animal rights and much was written on the topic. It wasn't until the twentieth century, however, that animal rights gained much greater prominence and public support. The increased use of animals in biomedical research, and the intensification in farming practices played a role in this. Perhaps one of the most critical events that renewed interest in animal rights, however, was the publication of a highly influential book by Peter Singer.

In 1975, Singer, a philosopher and applied ethicist from Princeton University, published *Animal Liberation*, in which he proposed that we should not discriminate between humans and animals based on species, because this promotes the view that we can ignore the interests of another animal just because they are not human. Singer does not believe that the interests of humans and animals should be given equal weight. It doesn't make sense for animals to have the same rights that we have; animals don't need freedom of speech or a right to vote, but Singer maintains that we should not ignore the needs of an animal just because it is not human. His philosophy and approach rely very much on the utilitarian ideas championed by Jeremy Bentham and others in previous centuries—how to balance an individual animal's needs with the consequences that might result by taking different actions. Interestingly, none of Singer's ideas are based on scientific data and fish are never mentioned.

For many, *Animal Liberation* provided a vision that they could identify with. These were members of society who firmly believed no animal should suffer at our expense. Singer's book was a powerful, intellectual catalyst that helped to give this part of society the momentum they needed to become an entity—'the animal rights movement' that we recognize today. Singer chooses to refer to it as 'the animal movement', carefully avoiding the use of the word 'rights'.

Since the 1970s the animal movement, has brought about numerous changes that have certainly improved the lives of many animals. The use of animals in product testing is now more regulated and where possible, alternatives to animals are found. While many in the animal movement would prefer society to become vegetarian, their campaigning has undoubtedly made us think about the welfare needs of the animals we rear for meat and other products. Cramped battery cages that don't allow chickens enough room to stretch their wings have been banned in Switzerland and Austria and are now being phased out across Europe. Even consumers have played a proactive role in this—where once the local supermarket may have filled shelves with eggs from battery-caged hens, those same shelves are now mostly stocked with eggs from free-range hens or hens kept in barns. The animal movement has changed the way that we protect and care for animals used in scientific research. Singer and colleagues would describe the animal movement as a work in progress and argue that there is still much to be done. But overall, there is now a much greater awareness about and sympathy for animal welfare, and discussions of our use of animals and the related ethical issues can now be heard in schools and read about in the popular press.

Running a similar yet different course to Singer's philosophy, the animal rights movement has also grown in the past thirty years, but not without controversy because some of the methods adopted by particular groups have been violent and illegal. Some supporters of animal rights believe that animals should be given legal rights and be considered to be part of the moral community. Animals should not be thought of as property and should not be used for food, as materials for clothing, or for research purposes. To draw attention to this view many animal rights groups decided it was necessary to take 'direct action'. These groups have specifically targeted the use of animals in biomedical research, and blood sports—such as fox and deer hunting.

In Britain, groups such as the Hunt Saboteurs Association, a non-violent organization, have used various methods to hinder the hunting of deer or foxes. The saboteurs blow horns and whistles to confuse the dogs, lay down false scent to throw the dogs off trails, and shut gates to make it difficult for hunters to keep up with the chase. Over the years their activities have caused considerable disruption, and this group played a key role in the recent banning of deer and fox hunting that came into effect in England and Wales in February 2005, and slightly earlier in Scotland. These protests against blood sports affected public opinion and ultimately led to changes in UK law. These kinds of action most likely explain why anglers have not been too interested to find out whether fish feel pain.

Angling has already been targeted by animal rights groups. Ad campaigns by groups such as People for the Ethical Treatment of Animals (PETA) have been run on billboards and the Internet. A few years ago PETA ran a campaign that showed a dog with a fishing hook pulling on it's upper lip with a slogan that read 'If you wouldn't do this to a dog, why do it to a fish?' This provocative advertisement was removed from a number of American billboards because it was deemed too sensitive a topic given that many states do very well from tourism associated with fishing. The advertisement is intended to shock people into overcoming the problem of the fish image that we discussed earlier—people regard fish as less important and so are less likely to react to a fish being hooked than say a fox being chased to exhaustion. More recently PETA has taken a different approach to try and convince people to rethink fish, in a campaign that suggests we should completely re-brand fish. Forget the old image of cold-blooded, slippery, and wet associated with the term 'fish'. They seek to rename fish as 'sea-kittens'. We wouldn't feel comfortable capturing a kitten with a hook so why a fish?

While the animal rights groups may be at one end of the spectrum, at the other end are people who refuse to believe that it is possible for fish to feel pain. The 2002 scientific article written by James Rose summarizes this stance. This review was published at least a year before the results of our experiments (described in Chapter 3) were released; yet the article continues to be used by various fishing societies and by many anglers as proof that fish cannot feel pain. The review is now seven years old, and the field has changed rapidly during that time. On discussing the topic with a number of anglers from several different countries it seems that opinions are now divided; while some continue to firmly dispute the capacity for fish to suffer from pain, others are beginning to reassess the situation and to reflect on how their actions could be detrimental to the fish.

We might have taken a long time to ask whether fish feel pain for many reasons. While accepting that it is a new area of enquiry, there is still much to be done. I believe that the weight of evidence now shows fish do feel pain. The next question we must address is what this means. This is the subject of the final chapter.

Looking to the Future

I have argued that there is as much evidence that fish feel pain and suffer as there is for birds and mammals—and more than there is for human neonates and preterm babies. In most developed countries, we treat chickens, pigs, cows, cats, and dogs differently because we believe they are sentient. If fish suffer, what are the implications for the way we interact with them? What changes can we make in the way we use fish to reduce their pain and suffering? Recognizing that fish have the capacity for pain perception has generated a desire among some to offer fish appropriate protection. If history is any guide, this desire will only get stronger. But while I agree that an aspiration for fish welfare is warranted by the evidence, we must be careful how we approach it and in particular we should be mindful that our understanding and the science underpinning what fish need or prefer is only just beginning. There is a worrying tendency at this point to rush at solutions; until we know more, we should be cautious in what becomes recommended or lawful practice.

One of the earliest groups to recognize the implications of pain perception in fish was the aquaculture industry. Many managers of fish farms are now interested in learning what can be done to improve the welfare of the fish they rear and how to decrease or change activities that lead to stress, pain, or suffering. Managers and industry stakeholders are taking this seriously for a number of reasons. Today, consumers take a greater interest in the sourcing of the food that they eat, and some are prepared to pay a little extra if this allows them to purchase animal products from recognized sources where the welfare of the animals has been taken into account. From a business perspective it makes sense to sell a product that consumers find desirable, and where the consumer is prepared to pay more, then there can be financial rewards for companies able to demonstrate that they operate with a high standard of care and welfare.

In fact fish welfare is something fish farmers have always strived for and the issues are quite black and white. Good welfare represents a win–win situation. A fish that is looked after will be healthy and will grow well, and attention to fish welfare helps promote large yields of pristine, good tasting fish. Poor welfare and neglect, on the other hand, can lead to outbreaks of disease, poor growth, and suboptimal flesh quality. Diseases can represent a considerable financial burden to the farmer because of the cost of the drugs required to treat the fish, and if treatment is not possible or fails, then whole cages containing thousands of fish sometimes must be destroyed. So good welfare is in the fish farmer's interest. Rearing fish en masse was a relatively new experience for those trying it out in the 1970s and 1980s, and at the time there was a steep learning curve with regard to discovering what worked and what didn't.

During the early stages when the aquaculture industry was rapidly expanding, mistakes were made and these were costly both in terms of direct losses and also in respect of the industry's image. High-density rearing led to outbreaks of infectious diseases that in some cases devastated not just the caged fish, but local wild fish populations too. The negative impact on local wildlife inhabiting areas close to the fish farms continues to be an ongoing public relations problem for the industry. Furthermore, a general lack of knowledge and insufficient care being taken when fish pens or cages were initially constructed, meant that pollution from excess feed and fish waste created huge barren underwater deserts. These were costly lessons to learn, but now stricter regulations are in place to ensure that fish pens are placed in sites where there is good water flow to remove fish wastes. This, in addition to other methods that decrease the overall amount of uneaten food, have helped aquaculture to clean up its act. With this history lurking in the background, it seems quite possible that the industry's enthusiasm to embrace fish welfare initiatives is a proactive effort to obtain a positive image.

Whatever the motivation, welfare is now on the fish farming agenda and several members of the aquaculture community have begun collaborating with fish scientists to identify which current practices adversely affect farmed fish. Researchers have already identified various routine handling practices that are stressful for the fish and in some cases this has led to changes in practice. Size-grading, for example, was once a very labour-intensive process requiring fish to be netted and handled, but now pumps and wide-diameter hoses fitted with counters are used to move fish of different size between tanks or ponds. This decreases both the amount of direct handling and the time fish spend out of water. The results are clearly beneficial for the fish: grading is now a less stressful process and the fish recover more quickly. The positive experience of improving grading procedures has led to an interest in finding other ways of improving routine aspects of husbandry.

Currently scientists are trying to determine what fish want when they are in captivity. Identifying preferences for different types of resource have had an impact in terrestrial farming studies of welfare. But one thing earlier experiments with terrestrial farm animals has shown is that 'preference tests' must be carefully designed and it is not always as straightforward as offering two or more choices and seeing what animals do. For instance, simple choice tests don't tell us how strongly the animal feels about its choice; in a bitter winter you might slightly prefer wool socks to cotton ones because they keep your feet a bit warmer, but you would probably strongly prefer to live sockless in a house with central heating than outdoors with excellent socks. Choice tests between two different types of sock or different types of housing would show that you prefer wool and central heating but they wouldn't reveal how strongly you felt about these choices. For assays in farms to be useful, we need to find ways of measuring how motivated an animal is to have access to one resource over another.

Marian Dawkins at Oxford University has pioneered a number of studies that explore what animals prefer and how much they want it. To do this she has designed experiments where animals must work for access to different choices. Measurements of how hard the animal will work begin to tell us how valuable that resource is to the animal. This provides a way of measuring the animal's strength of preference. For example, animals can be trained to push against a door to open it to get access to one type of resource or another. By putting heavier and heavier weights onto the doors you can find out how hard the animal is prepared to work to reach its choice. Similarly you can make the animal do something it prefers not to. Chickens do not like to squeeze through narrow spaces, but they will if this gives them access to something they really desire such as a place where they can dust bathe.

These kinds of approach can certainly be adopted to learn how fish value different types of housing or how important enrichment might be within a cage. Enrichment, such as objects and structures within the cage that the fish can interact with and manipulate, has turned out to be important for captive cod. In the wild, cod naturally spend much of their time close to the seafloor manipulating kelp and other things with their mouths. In cages on a farm there are many fewer opportunities for the fish to do this, but the cod found that they could fulfill this motivation by biting on the netting walls of the cages. This led to some early disasters because cod chewed away until holes appeared allowing the fish to escape. Work is currently underway to investigate what kinds of object the cod can be encouraged to chew and manipulate rather than the cage walls. Such problems never arose in salmon because they don't have the same desire to chew, and as salmon would naturally swim in mid-water without objects around them to interact with, the need for enrichment has never arisen on salmon farms. Differences in habitat use and behaviour are important to consider when designing ways of housing fish.

Other studies have explored how fish prefer to be fed. Originally feeding was a fairly rudimentary process where a farm worker would visit the cages and would scoop up processed feed pellets with a shovel and then fling these out across the pen. The area at the surface of the pen would appear to boil as the fish crushed themselves into the surface waters, scrambling with one another for access to food. Other early systems involved automated feeders that would similarly shoot and spray food pellets across the top of a pen at set times of day. The result of mass competition at the water surface was the same: aggression such as biting and chasing as the fish competed for access to food. Seeing the problems associated with the scramble competition, researchers tried to devise ways of letting the fish choose when to feed. This has now led to a completely different form of feeding. Fish operate the feeders themselves, nudging a panel with their snout or pulling on a string to release a small meal of food pellets. This method has several advantages. It puts the fish in control of its own feeding pattern, which has led to better growth rates through reduced stress at least in part. Furthermore, the fish use less food when fed this way, which saves the farmer money and less waste is generated, which is better for the surrounding environment.

Well-designed preference tests are a practical way to improve best practice for fish farms, but keeping large numbers of fish in confined areas can generate a variety of problems that often go beyond determining what fish want. A recurring issue faced by salmon and trout farms is damage that occurs to certain fins, in particular the tail and also the front pair of fins by the gill covers known as pectoral fins. When you buy whole salmon or trout you can almost always tell whether it is farmed or wild by the state of its fins. Wild-caught fish have large well-shaped, smooth-edged fins, whereas farmed fish typically have stubby, eroded, often ragged fins. The initial causes of fin trouble are still contested. Some think it may be a result of physical damage as the fins come into contact with the net walls of the cages or other rough surfaces, but others suggest it arises because of fish aggressively nipping at each other. Regardless, the consequences of bad fin damage can be disastrous; there are times when secondary bacterial infections become so bad that whole cohorts of fish need to be sacrificed. Previous work by the Russian biologist Professor Chervova, at the State University in Moscow, has shown that most fins are sensitive to being punctured and to pinching, and although nociceptors associated with the fins are yet to be described, there are nerve bundles within the fins, suggesting nociception is likely. Damage to fins may therefore be a painful experience. Moreover, when salmon and trout are aggressive with each other they often intentionally nip at each other's fins. This behaviour is understandable if the fins are sensitive areas and the fish experience such nips as painful.

Scientists and fish farmers are also currently targeting several other welfare-related problems such as the effects of water quality and trying to determine the optimal stocking density—to what extent can fish be crowded in tanks,

ponds or cages? This last question may seem straightforward but the answer turns out to be remarkably complicated
and variable. Advisory bodies that oversee the welfare
standards on fish farms and in research facilities have tried
to produce guidelines and legislation on how many fish
can be kept in a certain volume of water. Inspectors like
the concept of animal density because it is a relatively clean
measure, unlike other aspects of welfare that are often
difficult to assess. It works well for lab mammals—there
are agreed numbers of mice and rats that can be kept in
cages of a given size. But finding the right stocking densities for fish has been an ongoing problem. The trouble is
that there doesn't seem to be a single, good answer. How
many fish can be kept in a cage or a tank varies with
multiple factors such as the general health of the fish, age,
the social interactions between the fish, the level of feed
available, the quality of the water, and not least the species
concerned. In a farm setting all of these factors can and do
vary, so the optimal stocking density also varies. So while
it appears to be an easy target for those concerned with
welfare, the reality on the farm is quite different, and what
may be a good stocking density in one situation may lead
to compromised welfare in others.

This difficulty has triggered alarm bells because those
overseeing appropriate welfare standards wish to formal-
ize what they believe to be suitable stocking densities. The
problem isn't just limited to aquaculture; it also affects the
fish we house in research facilities and public aquaria, and

it is beginning to affect recreational fishing. Certain kinds of management programmes involve stocking hatchery-reared fish into streams, but how many fish should be released into different sections of a stream or river? There are concerns that our desire to act ethically and to provide fish welfare is ahead of our current scientific understanding. We need to proceed cautiously at this point and not rush through uncertain policies or inappropriate legislation.

A very real problem faced by those trying to devise guidelines and protocols to promote the welfare of captive fish is the sheer breadth of species that we use—this diversity makes it tricky to create generic guidelines. Different species have different requirements; they behave in different ways, they have different specialized sensory systems, and some are better at coping with the captive environment than others. Furthermore, fish are much more variable than their terrestrial cousins. Traits fixed in mammals and birds are often more changeable in fish. Several species of fish have the ability to change sex, for instance, and not just once. There are species in which individual fish literally shuttle back and forth between male and female, depending on the nature of the current social environment. Other changeable characteristics are jaw morphology and gill structure. These can rapidly change as fish specialize at feeding on certain types of prey—final snout shape can depend on which food types were abundant as fish developed.

Other less labile features also make fish different. Almost all fish are dependent on the environment around them for their body temperature. Their physiology requires less food than warm-blooded animals of a similar size. Certain species of fish, such as salmon and trout, even choose not to eat for long periods of time, so food deprivation may be less of a challenge to fish than it is for birds and mammals. Such differences between species and their general plasticity make it next to impossible to create blanket instructions for fish, analogous to those used for rats or mice in research labs. The Council of Europe has begun to tackle this by approaching fish specialists and inviting them to prepare species-specific information sheets. It may take some time to gather these and to validate them, but once they have been prepared they will be an excellent resource. Furthermore, they will provide more appropriate, tailored care instructions to help us consider the needs of the fish that we hold in captivity. In the meantime, we must be willing to accept that there are few simple solutions.

As we try to improve the welfare of captive fish, questions about other ways we interact with fish, particularly with regard to angling and sport fishing, arise. The effect of the fish pain debate with regard to angling is at an early stage, and what the future holds for recreational fishing is by no means certain. In the last chapter the animal activist issue was discussed and it seems certain that their views and opinions will become a prominent part of the fish pain debate. To counter the criticism of activists, a number of fish biologists who are also keen anglers have already begun investigating how certain routine practices connected with angling affect the fish that have been caught. These kinds of research are an important contribution to our general understanding of fish—how capture time, handling time, and exposure to air affect stress responses and the conse-quences of keeping the fish confined in keepnets once unhooked. The results are already having an impact on the equipment and gear anglers choose to use and how anglers interact with the fish they catch. What we learn from these studies can be implemented into codes of practice to guide anglers about what is and isn't good for the fish. In the same way that those overseeing welfare standards on fish farms need scientific evidence to underpin their decisions about best practice, so our understanding of fish behaviour and their responses during angling interactions needs to be grounded in carefully designed research. These studies are being undertaken and the results are being disseminated and listened to because the majority of anglers care about the fish they catch.

There are a number of myths and misconceptions that need to be addressed. Two examples of misconceptions frequently used in arguments against the ability for fish to feel pain are worthy of discussion here. First, it is often suggested that because hooked fish swim and pull away from the angler—causing more pressure in the flesh around the hook—the hook doesn't adversely affect the fish or truly hurt it. Second, if fish are smarter than we have previously given them credit for and biting a hook really is a painful experience then why is it possible to hook the same fish again and again? The fact that they are not learning to avoid hooks would suggest that the hooking process isn't very unpleasant.

Let me tackle these two misconceptions in turn. When vertebrates are trapped or caught, their body expresses a number of responses, often without the need for conscious thought, rather like the nociceptive part of the pain response. The sympathetic nervous system controls the fight-or-flight response, preparing the body for an acute stress response to enable the animal to either fight or flee in a threatening situation. The hook may well cause the fish pain, but as with other animals in difficult situations, the motivation to escape is so strong that the fish works to overcome any pain to try and get away. We sometimes hear of similar reports in other animals, including ourselves. People who have lost part of a limb in an accident can sometimes complete what seem to be extraordinary heroic acts before collapsing in an ambulance or a hospital bed. Animals caught in leg-traps sometimes gnaw off their own leg, presumably causing themselves considerable pain, in order to escape. In 2003 a hiker sawed his own arm off with a pen-knife to escape from a fallen bolder, and 10 years before this, an angler cut his leg off at the knee when two rocks shifted trapping him in a remote river.

There is some evidence trout can also change the way they respond to pain. Lynne Sneddon recently investigated how trout vary their response to painful stimuli under different social conditions. Fish given a painful stimulus when they perceived themselves to be in a dominant social position continued to express dominant, aggressive behaviour presumably because maintaining dominant status had a higher priority than expressing signs of pain. The escape behaviour of a fish hooked on a line is an example where competing motivations influence fish behaviour—the motivation to escape being higher than the motivation to respond to the hook through the jaw. We would expect evolution to have promoted such decision-making processes: pain in the mouth is nothing compared to the loss of life. Thus in very threatening situations the motivations of an animal come under the influence of the fight-or-flight response and this can mask the usual responses made to certain kinds of stimuli.

It is true that fish can be caught more than once and sometimes scars from previous hooking events can be seen on the jaws of fish. But this does not mean fish are indifferent to the process of being hooked; fish in some circumstances can learn to avoid artificial flies and lures. Several years ago Professor Jan Beukema from the Netherlands Institute for Sea Research investigated the response of pike being caught on a rod and line. Using tagged individuals so that he could keep track of which individuals were caught, his study showed that pike quickly learned to avoid hooks after just one day of exposure to fishing. In a separate study, again with a captive population of fish in a pond, Beukema reported that carp initially caught by a rod and line would later need to be recaptured employing a different technique such as directly netting the fish because their initial experience made them wary of the rod and line. Other anglers have also reported that in places where fish are typically released after being caught, trout learn to stop taking artificial bait if fishing pressure is high. So why don't all fish learn? In a wild stream where there is plenty of competition around for a limited supply of food it seems likely that many fish will not be able to afford the luxury of being choosy about what they attempt to eat. The motivation to feed and the competition from nearby neighbours are likely to be strong, which may lead to fish making mistakes and taking the bait on more than one occasion. Beukema's experiments with pike and carp in which fish avoided being hooked a second time were run in an enclosed pond where the fish were well fed; their hunger levels were presumably much lower than that of the average fish in a stream or river.

The fact that in certain circumstances fish do learn to avoid being hooked does imply that they find it an unpleasant process. This raises questions about the ethics of a now common form of recreational fishing—'catch and release'. The release of the fish is a fisheries management strategy used to conserve the animals within the system. This approach, however, is not without controversy because it contradicts the general philosophy and ethics underlying the capture of wild animals where a 'clean kill' after capture is deemed to be best practice. A similar philosophy is found in the UK Animal Scientific Procedures Act that protect animals used in research—the use of an animal in more than one experiment is banned unless a very clear case can be made to justify the need to test the same animal in different circumstances. These kinds of experiment are very rare and reuse is avoided wherever possible; the aim is to avoid repeatedly stressing individual animals. From an ethical perspective, we accept that certain procedures we undertake may cause a degree of pain or suffering and this should be minimized as much as possible. The problem with reusing an animal is that it may be further exposed to potential suffering. What does this mean then for the fish in a catch-and-release fishery? This is certainly one area where input from bioethicists would be helpful. Is catch and release acceptable if it is part of a population management process? Is it appropriate to put an individual fish through the experience of being caught by a hook more than once?

Catch and release is considered by many cultures to be 'unnatural'. For these people, fishing is all about going out to catch food. The fish you land will be used for a meal, so catching a fish only to put it back again makes no sense to such people and is sometimes considered to be 'playing with food'. Others also believe catch and release to be ethically wrong, and in some cases this has led to changes in legislation that affect recreational fishing. In Germany for example, laws forbidding the intentional release of fish over a certain size have now been passed; in other words, you must kill and remove any fish you catch if it is large enough. For ethical reasons, you are not allowed to put these individuals back into the river or pond. A similar ruling was also recently introduced in Switzerland. There is great concern among those who choose to fish for recreation about what these kinds of ruling will do to their pastime. The logic underpinning the new legislation has also been queried. For instance, is taking the largest fish the best idea? Large females will be able to lay more eggs by virtue of their larger size, so from a management perspective these fish might be the ones that would help you maintain reasonable numbers of fish within the population.

My view on this particular issue is that, like the stocking density problem in fish farming, we may be running ahead of ourselves and creating legislation before we are ready. Catch and release is an ethically difficult issue; knowing that fish feel pain and can suffer raises questions about whether it is appropriate to allow fish to be caught multiple times. But if catch and release is banned, what do we do in rivers where there aren't enough fish to support the pressure of angling? Should these rivers be intentionally stocked with hatchery fish, many of whom die from starvation or predation because the predator-free, food-rich hatchery environment does little to prepare these fish for life in a natural river? What is better, accepting the negative effects of recapturing a few fish multiple times or the one-off effects of capturing many more fish just once? This is clearly another area where bioethicists are needed to help us navigate our way to an informed, ethically appropriate solution. Is there an ethical case to be made that will permit recreational angling that involves catch and release? If not, can recreational angling continue to exist? What would be the consequences of banning angling? It may not improve fish welfare. Many current aquatic conservation projects and efforts that go to maintaining good, clean waterways are championed by anglers. If recreational angling is no longer encouraged, concerns are expressed about what will become of rivers and ponds and the numerous non-target wildlife species that currently benefit by the presence of anglers.

If we are to undertake an ethical analysis of recreational angling we need to know how the angling process affects fish. The research that fish biologists have been doing can help here. Their findings are already having a direct effect on what is now regarded to be best practice and this has had an impact on what young anglers are taught. As mentioned earlier, practices such as removing the barb from a hook prior to use are now promoted—getting rid of the barb makes it easier to more cleanly remove the hook from the fish after it is caught. Other commonly taught techniques are using wet rather than dry hands to handle the fish and keeping fish submerged in water as much as possible.

Anglers are increasingly encouraged to think about the choices that they make with regards to equipment and gear. Knotless nylon or rubber nets, for example, are less abrasive and do less damage to the body of the fish. Knotted nets, on the other hand, can cause skin abrasions, which may be mild at the time of capture but which leave the fish prone to later fungal infections. The right choice of shape and size of hook is important; too big a hook can inflict unnecessary tissue damage compared to a smaller hook. When choosing which shape and size of hook to use, the gape of the mouth and general head and jaw shape of the species being fished should be considered. Hooks should always be removed from fish that are to be released, and if a hook has become stuck and cannot be removed without causing significant tissue damage then the fish should be killed because the presence of the hook can lead to infections and make it difficult for the fish to feed. The physical environment can also have an effect on how well the fish cope with being caught and handled. Water temperature influences the physiology of the fish and warmer temperatures can lead to increased risk of death in fish held in keep-nets.

The knowledge we have gained about how fish respond to different types of equipment and environments allows the modern recreational angler to take greater care as they interact with the fish they catch. As discussed earlier, many anglers fish because they are fascinated by fish and far from wanting to do them harm they are keen to ensure the careful handling and release of the individuals that they catch. The majority of anglers feel very passionate about their pastime and are very keen to protect recreational fishing from those who would push for this pursuit to be banned. And indeed, many good things can come from angling: it fosters an interest in nature and wildlife, and it motivates people to care for the natural environment and to guard against pollution. It encourages younger members of society to go outdoors and temporarily leave their seemingly ubiquitous games consoles. Do these benefits to people and society outweigh the potential suffering that may be experienced by the fish? My view is that even accepting that fish feel pain does not make these calculations simple.

While angling is one way many people interact with fish, there are several other ways that fish are used. There is a growing trend for fish to be part of visual displays in public places such as shopping centers or restaurants; waiting rooms in doctors' surgeries or hospital outpatient areas increasingly incorporate displays of small tanks with various fish species. They are often considered to add a calming influence. Is it ethical to have fish on display in shopping malls or doctors' surgeries? Given the impressive moving images that can be displayed on television screens or computer monitors there are potential alternatives, but whether such substitutes have the same calming influence as the real thing has yet to be investigated. On a larger scale, public aquaria are regarded as valuable revenue-generating tourist attractions. The

welfare of the fish we house and maintain in these facilities is beginning to be queried. In comparison to other
areas where we interact with fish, there is almost nothing
known about the effects of captivity and how the fish
cope with brightly lit areas crowded, noisy, and bustling
with activity. A number of studies have considered the
welfare and ethics of zoo animals, and these arguments
will relate to captive fish in public aquaria. The sourcing
of fish for such displays is a welfare concern as the
majority of fish on display are wild caught and are not
reared in captivity. Thus issues of supply and transport
arise in addition to how fish cope with being moved from
the wild to captivity.

In terrestrial zoo animals, like tigers, elephants, and
polar bears, boredom, frustration, and enclosures too
small and too plain sometimes lead to 'stereotypies'—
repetitive actions or movements performed over and over
again. Likewise, sharks and other fish species that typically have large home ranges, or make long distance migra-
tions, also show stereotypies in public aquaria. These
behaviours are not necessarily painful but they do represent a welfare concern because they are expressions of
frustration on the animal's part, and certain repetitive
actions that involve contact with walls or other structures
can lead to the fish rubbing and damaging their skin tissue.
Finding ways to enrich and add variety to an environment
appropriately, can help alleviate boredom in zoo enclosures. This has yet to be tried for fish in aquaria. Careful

choice of which species to house in public aquaria, and avoiding species with known large home ranges would probably help to prevent the development of stereotypies.

Whether it is ethical to house animals in zoos simply for our entertainment has been debated many times. Over recent years there has been a shift in the justification for zoos, with more recent arguments focusing on the role of zoos in conservation projects. Changes in the natural environment through climate change, overexploitation, and anthropogenic disturbance are taking their toll on many different fish populations. Public aquaria have barely been previously considered as conservation tools, but this may be a role to which they could contribute to in the future. We are becoming increasingly skilled in breeding various species of fish in captivity, a beneficial spin-off that aquaculture has provided, but the diversity of fish and their many different reproductive strategies mean there is still a lot to learn before we could consider aquaria to be significant contributors to some sort of biological ark.

Perhaps the biggest negative effect we have on fish, and the least researched, is how we fish at sea. The welfare of marine fishes has been largely ignored, but harvesting fish in the open ocean has enormous potential to cause pain and suffering and not just for the fish themselves, but also for other creatures caught in the fishing nets. There are many different forms of fishing at sea: trawling, trolling, purse seining, gill netting, and long-lining. Long-line fishing catches species such as tuna, swordfish, and mahi mahi. Long-lining crews set up several hundred lengths of line that can be tens of metres long. The lines, rigged with floats and hundreds of baited hooks, are left for several hours at a specific depth in the water to attract hungry large fish species. Once hooked, depending on the number of lines set, the fish may have up to 10 hours to wait before the lines are collected in. Many fish are exhausted from trying to escape, but they are still alive as they are hauled onto the deck of the fishing vessel and then left to suffocate in the air.

This form of fishing has received a great deal of media attention not for fish welfare reasons but because the squid-baited hooks attract sea birds such as albatross who themselves then become caught on the hooks and drown. Hundreds of thousands of birds have died this way and four species of albatross are currently perilously close to extinction. Advertisements and appeals made by conservation groups campaigning against this form of fishing typically show pictures of bedraggled, drowned birds with hooks through their bills. These are remarkably reminiscent of the PETA 'You wouldn't do this to a dog' campaign picture of a hook through a dog's lip. The campaigns by bird conservation groups have successfully reduced the risk to birds, so that now fishing often occurs at night when fewer birds are out foraging, and streamers are attached to the lines to scare birds away. Hooks are checked more frequently, and weights are used to sink the lines beyond the diving depth of the albatross. Meanwhile, the baited hooks continue to catch fish, although one benefit of these changes for the fish is that the lines are now checked more frequently, decreasing the time from hooking to being hauled on deck.

Deep water long-lining which uses heavier, shorter main lines with side lines spaced out at roughly five-metre intervals operates several hundred metres down at the seafloor. These bottom long-lines have many more hooks attached to them than the shallower long-lines. The goal is to attract species such as cod, haddock, halibut, grouper, sea bream, and at much greater depths Patagonian toothfish. Deep water long-lining too has had its share of bad press in the past few years because of their impact on other creatures, particularly turtles, dolphins, and whales that are caught and then drown. Compared with other forms of fishing, however, this long-lining is actually quite targeted, primarily catching the species the fishing vessel is trying to hook. Other forms of fishing are not so selective.

Trawling, for instance, involves a large cone-shaped net towed behind one or a pair of vessels. The net has a wide opening at its mouth with large mesh size that leads back towards a closed end section with a much finer mesh that is eventually tied with a knot. The shape and design of the net funnels the fish towards the knotted end. Trawls can be set at different depths in mid-water to catch schools of fish, or they can be set deeper to catch bottom-dwelling fish. Fish caught in the net try to outswim it initially but the speed of the trawl vessels is regulated to avoid this.

Fish caught early on in the trawl eventually become exhausted from constantly trying to avoid the moving net and then drift in towards the knotted end where they are then pulled along within the net. At the end of the trawl cycle, motorized winches pull the net to the surface and then up over the deck.

As the trawl net moves up through the water column the rapid changes in pressure cause problems for the fish. When scuba divers come back to the surface after a dive has finished they must stop at different depths and give their bodies time to adjust to the changes in pressure; coming up too quickly causes barometric trauma. This is more commonly known as 'The Bends' because the severe joint pain that accompanies the condition causes people to double up as they try to cope with the pain. The problems arise because the rapid decrease in pressure causes nitrogen in the blood and tissues to come out of solution and form gas bubbles inside the body. This can be extremely painful and runs the risk of causing embolisms. Fish trapped in a net that moves up rapidly to the surface experience similar effects. For them, one of the bigger issues is their swim bladder. Without time to adjust to the decreasing pressure, the gas-filled swim bladder typically becomes overinflated, causing huge distention inside the fish. Sometimes the pressure is so great their stomach and intestines are pushed out of their mouth and anus. Eyes can also become distorted and bulge out.

Once the trawl net reaches the surface the fish begin to experience crushing effects because they no longer have the buoyancy provided by the water to counteract gravity. As the net is lifted from the water the fish towards the bottom of the net are literally squashed and compressed by the weight of fish above them. With the net finally positioned over the deck of the fishing vessel the knot is untied and the fish spill out. Still flapping, they are sorted into fish to be kept versus fish that are by-catch. Those that are still alive at this stage slowly suffocate in air—a process that can take several minutes. By-catch are often washed or thrown overboard, but the majority of these fish are in such bad shape that by the time they are discarded back into the sea few are capable of surviving. Concerns over the loss of animals caught as by-catch has forced the fishing industry to change the design of fishing gear. Trawl nets can now be fitted with devices such as turtle excluders—these, and other modifications to the net, allow by-catch animals to escape from the moving trawl before it is pulled in.

Whilst it is important to remember that until the fish are caught they have lived and grown in a natural environment, how different fishing methods impact on fish welfare is not known. My view is that based on the data we have for other fishes (discussed in Chapters 3 and 4), the balance of probabilities is that these marine fish will be experiencing pain and suffering as a result of their capture. Consequently, I think that just as we concern ourselves with welfare in other food production contexts, we should consider whether there are ways we can alleviate this suffering in marine fisheries. There are, and some action is being taken by certain parts of the fishing industry, but perhaps not surprisingly it is for economic rather than welfare reasons. Studies in Scotland have shown that several common harvesting practices damage the quality and price of the fish. For example, putting fish into boxes and group weighing them can damage the flesh and so the quality and value of the filet. It has been shown that individually weighed fish are worth twenty-five per cent more than box-weighed fish. Although this kind of handling happens after death, the message is sinking in: better handling and decreasing the stress of the fish at harvest produces a better product with a higher economic value. Larger crews have also been found to harvest a more valuable catch—not just in terms of the amount caught but in the quality of the fish brought to shore. More pairs of hands take better care of the fish as they are being landed and handled. As more and more fishing crews recognize that they can increase the profits they make by changing their routine practices, it may be possible to devise better, more humane ways to harvest fish at sea.

Certain markets such as the Japanese sushi market demand quickly killed fish to enhance the flavour and quality of the flesh. Iki Jimi, which literally means 'quick kill' in Japanese, is increasingly used on large, valuable fish destined for the sushi market. A spike is used to destroy the brain and to rapidly stop the fish from moving, once out of water. Culturally, this is considered a humane way of killing as it is very fast and quickly shuts down the fish's nervous system. For large fish such as those caught on surface long-lines this could be introduced as a more humane means of slaughter than letting the fish slowly suffocate in the air, but killing each fish as it is brought onto the boat is labour-intensive and will demand larger crews, which will force the price of the fish to increase. You could argue that it would be cheaper to drown chickens than individually kill them, but we don't do this.

The aquaculture industry has recently had to take a long, hard look at slaughter methods because some of the earlier techniques were poorly conceived and caused unwanted side effects; high stress levels in the fish were found to affect flesh quality. This led to questions about whether fish were actually dead at the point at which they were being processed. Carbon dioxide is used to slaughter terrestrial food animals such as chicken and pigs; it allows animals to be killed in batches and there is a general perception that the animals do not find it aversive or stressful, although recent research is calling that interpretation into question. As carbon dioxide is soluble in water it has also been used as a slaughter method in aquaculture. Fish are pumped out of their pens and transferred to tanks saturated with carbon dioxide. They are left in the water until they stop moving; at this point they are considered unconscious and are then removed from the tank. Their gills are then cut with a knife so that the fish bleed out before they are moved to tables where they can be processed. To keep the fish fresh they are often laid on or surrounded by crushed ice during the bleeding-out process. Using this method a relatively small team of people can kill a large number of fish.

Research has now shown that carbon dioxide followed by cutting the gills is actually not a good method of slaughter with regard to fish welfare because it leads to considerable stress and suffering. As the fish enter the CO_2-saturated water, their environment suddenly becomes very acidic, which irritates softer tissues such as the gills. The stress of the rapid change in environment often causes the fish to become very agitated and to excrete ammonia, further affecting the quality of the water within the tank. The fish struggle for several minutes before they become immobile from exhaustion and lack of oxygen. There is actually no evidence that the fish are anaesthetized at this stage—so they are not unconscious when their gills are cut. The ice that they are then packed into chills them, but because these animals are cold blooded, their metabolism simply slows, thus prolonging the time it takes them to die. While this technique may seem efficient because many fish can be killed at a time, it is no longer deemed a humane method of slaughter.

A desire to minimize the stress and suffering of farmed fish at harvest has led to the phasing out of carbon dioxide in Europe. Several researchers have developed other methods of which two, electrical and percussive stunning, are preferred. Electrical stunning involves passing a current through a small tank into which a few fish have been pumped, or the fish are pumped out onto a moving conveyor belt that is electrified. The exposure to the electric current instantaneously causes unconsciousness. The gills are then cut and the fish are allowed to bleed out before being chilled and processed. Percussive stunning involves literally knocking the fish unconscious. Mechanized systems have been devised where a few fish at a time swim down through channels that become narrower until the fish reach an end point. As soon as they reach this area an automated blow can be delivered to the head immediately, rendering the fish unconscious. With only a few fish stunned at a time, the workers can cut the gills and then process each batch before the next fish are allowed to enter the channels. Both electrical and percussive stunning has been shown to induce much less physiological stress and ensures unconsciousness before processing.

It has taken several years to develop these more humane slaughter techniques for fish farms, but they are slowly replacing older harvesting methods like carbon dioxide. Whether any of the methods developed in aquaculture could be used for fish caught at sea is not yet clear, but issues of how we handle and treat the fish we catch in the oceans are beginning to receive attention. For example the 2008 World Fisheries Congress held in Japan had a special session sponsored by the Fisheries Society of the British Isles at which they addressed fish welfare in wild-capture marine fisheries. The conclusion from this meeting was that there is a clear need for scientific evidence of what happens to fish as they are caught at sea. We have already seen some likely points of concern in the discussion above, which now need to be more formally assessed. With that knowledge we could begin looking for procedures that would lower the detrimental impact we have on fish. These will almost certainly be more labour-intensive and less efficient than current methods so the cost of fish for the consumer will increase. But if more humane harvesting methods can be found, then aren't we ethically obliged to invest in these? In large part, that has been the conclusion for terrestrial animal industries.

The experiments and results presented in the first part of the book provide enough evidence to answer the question posed in the title—'Do fish feel pain?' Yes, they do. As demands for good animal welfare increase, this answer will change the way we think and act. Some will struggle with this change. Fish are still perceived as 'different', but the deeper we delve, the more we recognize and appreciate many similarities with birds and mammals. The fish pain debate is gaining momentum. To advance the discussion we must review the facts dispassionately. Rational well-argued evidence, not intimidation, will be the way to make progress. It appears as though we are at a crossroads, and a number of options lie ahead of us. In choosing which of those routes to take we must understand that there is a lot more to learn about what fish need to promote their welfare. We need to address the ethics underlying our interactions with fish. Most importantly, we must proceed carefully with creating laws and guidelines, making sure we do so in an informed way. Accepting that fish experience pain and suffering does force us to think differently, and it will in due course force us to act differently in many spheres. But what that action should be is for the most part still unclear. Knowledge, education, and open minds are surely our best guides through this uncharted territory.

Bibliography

Suggested further reading based on the separate chapters, where possible reviews that summarise a topic are listed.

CHAPTER 1

Animals (Scientific Procedures) Act 1986, available at http://www.archive.official-documents.co.uk/document/hoc/321/321-xa.htm. Accessed November 2008.

Dawkins, M. S., *Through Our Eyes Only? The Search for Animal Consciousness* (Oxford: Spectrum/Freeman, 1993).

Farm Animal Welfare Council: Five Freedoms, available at http://www.fawc.org.uk/freedoms.htm. Accessed April 2009.

Montgomery, J. C., and A. G. Carton, 'The Senses of Fish: Chemosensory, Visual and Octavolateralis', in C. Magnhagen, V. A. Braithwaite, E. Forsgren, and B. G. Kapoor (eds), *Fish Behaviour* (Enfield, NH: Science Publishers, 2008), 3–31.

Nagel, T., 'What Is it Like to Be a Bat?', *Philosophical Review* 83 (1974), 435–50.

Russell, W. M. S., and R. L. Burch, *The Principles of Humane Experimental Technique* (London: Methuen, 1959; rpt Hertfordshire: Universities Federation for Animal Welfare, 1992).

Sneddon, L. U., V. A. Braithwaite, and M. J. Gentle, 'Do Fish Have Nociceptors? Evidence for the Evolution of a Vertebrate Sensory System', *Proceedings of the Royal Society, London Series B* 270 (2003), 1115–21.

Vallortigara, G., and A. Bisazza, 'How Ancient is Brain Lateralization?', in L. J. Rogers and R. J. Andrew (eds), *Comparative Vertebrate Lateralization* (Cambridge: Cambridge University Press, 2002), 9–69.

CHAPTER 2

Barr, S., P. R. Laming, J. T. A. Dick, and R. W. Elwood, 'Nociception or Pain in a Decapod Crustacean?', *Animal Behaviour* 75 (2008), 745–51.

Bateson, P., 'Assessment of Pain in Animals', *Animal Behaviour* 42 (1991), 827–39.

Brace, R. C., and S.-J. Santer, 'Experimental Habituation of Aggression in the Sea Anemone *Actinia equina*', *Hydrobiologia* 216 (1991), 533–7.

Fitzgerald, M., 'The Development of Nociceptive Circuits', *Nature Reviews in Neuroscience* 6 (2005), 507–20.

Kavaliers, M., 'Evolutionary and Comparative Aspects of Nociception', *Brain Research Bulletin* 21 (1988), 923–31.

Qui, J., 'Does It Hurt?', *Nature* 444 (2006), 143–5.

Weary, D. M., L. Neil, F. C. Flower, and D. Fraser, 'Identifying and Preventing Pain in Animals', *Applied Animal Behaviour Science* 100 (2006), 64–76.

CHAPTER 3

Chervova, L. S., 'Pain Sensitivity and Behaviour of Fishes', *Journal of Ichthyology* 37 (1997), 106–11.

——, and D. N. Lapshin, 'Opioid Modulation of Pain Threshold in Fish', *Doklady Biological Sciences* 375 (2000), 590–1.

Dunlop, R., and P. Laming, 'Mechanoreceptive and Nociceptive Responses in the Central Nervous System of Goldfish (*Carassius auratus*) and Trout (*Oncorhynchus mykiss*)', *Journal of Pain* 6 (2005), 561–8.

——, S. Millsopp, and P. Laming, 'Avoidance Learning in Goldfish (*Carassius auratus*) and Trout (*Oncorhynchus mykiss*) and Implications for Pain Perception', *Applied Animal Behaviour Science* 97 (2006), 255–71.

Nordgreen, J., T. E. Horsberg, B. Ranheim, and A. C. N. Chen, 'Somatosensory Evoked Potentials in the Telencephalon of Atlantic Salmon (*Salmo salar*) Following Galvanic Stimulation of the Tail', *Journal of Comparative Physiology A* 193 (2007), 1235–42.

Sneddon, L. U., V. A. Braithwaite, and M. J. Gentle, 'Do Fish Have Nociceptors? Evidence for the Evolution of a Vertebrate Sensory System', *Proceedings of the Royal Society, London Series B* 270 (2003), 1115–21.

——, ——, and ——, 'Novel Object Test: Examining Pain and Fear in the Rainbow Trout', *Journal of Pain* 4 (2003), 431–40.

Yue, S., R. D. Moccia, and I. J. H. Duncan, 'Investigating Fear in Domestic Rainbow Trout, *Oncorhynchus mykiss*, Using an Avoidance Learning Task', *Applied Animal Behaviour Science* 87 (2004), 343–54.

CHAPTER 4

Dawkins, M. S., 'Who Needs Consciousness?', *Animal Welfare* 10 (2001), 519–29.

Chandroo, K. P., I. J. H. Duncan, and R. D. Moccia, 'Can Fish Suffer? Perspectives on Sentience, Pain, Fear and Suffering', *Applied Animal Behaviour Science* 86 (2004), 225–50.

——, S. Yue, and ——, 'An Evaluation of Current Perspectives on Consciousness and Pain in Fishes', *Fish and Fisheries* 5 (2004), 1–15.

Mendl, M., and E. S. Paul, 'Consciousness Emotion and Animal Welfare: Insights from Cognitive Science', *Animal Welfare* 13 (2004), 17–25.

Rodríguez, F., C. Broglio, E. Dúran, A. Gómez, and C. Salas C, 'Neural Mechanisms of Learning in Teleost Fish', in C. Brown, K. Laland, and J. Krause (eds), *Fish Cognition and Behaviour* (Oxford: Blackwell, 2006), 243–77.

Rolls, E. T., *Emotion Explained* (Oxford: Oxford University Press, 2007).

Rose, J. D., 'The Neurobehavioral Nature of Fishes and the Question of Awareness and Pain', *Reviews in Fisheries Science* 10 (2002), 1–38.

——, 'Anthropomorphism and "Mental Welfare" of Fishes', *Diseases of Aquatic Organisms* 75 (2007), 139–54.

Salas, C., C. Broglio, E. Dúran, A. Gómez, F. M. Ocana, F. Jimenez-Moya, and F. Rodriguez, 'Neuropsychology of Learning and Memory in Teleost Fish', *Zebrafish* 3 (2006), 157–71.

CHAPTER 5

Appel, M., and R. W. Elwood, 'Motivational Trade-offs and the Potential for Pain Experience in Hermit Crabs', *Applied Animal Behaviour Science* 119 (2009), 120–4.

Clayton, N. S., and A. Dickinson, 'Episodic-Like Memory during Cache Recovery by Scrub-Jays', *Nature* 395 (1998), 272–4.

Elwood, R. W., and M. Appel, 'Pain Experience in Hermit Crabs?', *Animal Behaviour* 77 (2009), 1243–6.

——, S. Barr, and L. Patterson, 'Pain and Stress in Crustaceans?', *Applied Animal Behaviour Science* 118 (2009), 128–36.

Emery, N. J., and N. S. Clayton, 'Effects of Experience and Social Context on Prospective Caching Strategies in Scrub-Jays', *Nature* 414 (2001), 443–6.

Hampton, R. R., 'Rhesus Monkeys Know When They Remember', *Proceedings of the National Academy of Science USA* 98 (2001), 5359–62.

Harding, E. J., E. S. Paul, and M. Mendl, 'Cognitive Bias and Effective State', *Nature* 427 (2004), 312.

Mather, J. A., 'Cephalopod Consciousness: Behavioural Evidence', *Diseases of Aquatic Organisms* 17 (2007), 37–48.

CHAPTER 6

Dawkins, M. S., 'Using Behaviour to Assess Animal Welfare', *Animal Welfare* 13 (2004), 3–7.

Clover, C., *The End of the Line* (London: Ebury Press, 2004).

Fraser, D., *Understanding Animal Welfare* (Chichester: Wiley-Blackwell, 2008).

People for the Ethical Treatment of Animals, available at http://www.peta.org. Accessed August 2009.

Singer, P., *Animal Liberation* (New York: HarperCollins, 1975, rpt 2001).

CHAPTER 7

Arlinghaus, R., S. J. Cooke, L. Lyman, D. Policansky, A. Schwab, C. Suski, S. G. Sutton, and E. B. Thorstad, 'Understanding the Complexity of Catch-and-Release in Recreational Fishing: An Integrative Synthesis of Global Knowledge from Historical, Ethical, Social, and Biological Perspectives', *Reviews in Fisheries Science* 15 (2007), 75–167.

Beukema, J. J., 'Angling Experiments with Carp (*Cyprinus carpio* L.). II. Decreasing Catchability Through One-Trial Learning', *Netherlands Journal of Zoology* 20 (1970), 81–92.

——, 'Acquired Hook Avoidance in Pike *Esox lucius* L. Fished with Artificial and Natural Baits', *Journal of Fish Biology* 2 (1970), 155–60.

Branson, E. J., *Fish Welfare* (Oxford: Blackwell, 2008).

Davie, P. S., and R. K. Kopf, 'Physiology, Behaviour and Welfare of Fish during Recreational Fishing and after Release', *New Zealand Veterinary Journal* 54 (2006), 161–72.

Hart, P. J. B. and Reynolds, J. D., *The handbook of Fish Biology and Fisheries*, Volumes 1 & 2 (Oxford, Blackwell, 2002).

Huntingford, F. A., C. E. Adams, V. A. Braithwaite, S. Kadri, T. G. Pottinger, P. Sandoe, and J. F. Turnbull, 'Current Understanding on Fish Welfare: A Broad Overview', *Journal of Fish Biology* 68 (2006), 332–72.

Jamieson, D., *Morality's Progress* (Oxford: Oxford University Press, 2002).

Policansky, D. 'The Good, Bad and Truly Ugly of Catch and Release: What Have We Learned?', *Wild Trout Symposium* 9 (2007), 195–201.

INDEX